Praise for *Feline Pi.*

"Engaging, amusing, perceptive and untimely, in the most admirable Nietzschean sense."　　　　—Mark Rowlands, *New Statesman*

"Cat lovers will enjoy the celebration of feline mythos, from the cat gods of ancient Egypt to purring contemporary domestics, while hardcore Gray fans will be reassured by the usual references to immortality cults, Hobbes, the gulags and so on."
　　　　　　　　—Charles Arrowsmith, *The Washington Post*

"Curious and exploratory. Gray moves freely among writing modes, . . . [telling] stories of famous cats, dabbling in evolutionary history and showing a clear appreciation for his subject. Above all, the book is an ode to cats, and Gray gives the impression of having learned from them how to take pleasure where he finds it."
　　　　　　　　　　　　　　　　—*Kirkus Reviews*

"A wonderful mixture of flippancy and profundity, astringency and tenderness, wit and lament."　　　　—Jane O'Grady, *The Telegraph*

"A scratching, spitting, and finally purring tour de force."
　　　　　　　　　　—Will Self, author of *Umbrella*

"Gray offers well-timed scratches at tender bits of the human psyche and sinks some sharp teeth into a few of our most cherished self-conceptions."　　　　—Ian Ground, *The Times Literary Supplement*

"Slyly playful . . . As enlightening as it is delightful."
　　　　　　　　　　—John Banville, *The Irish Times*

"Magnificent."　　　　　　　　—Kathryn Hughes, *Literary Review*

"Gray's work makes a strong case that our species is incorrigibly irrational, and it raises questions about humanist beliefs that should be particularly important for those of us on the political left to consider . . . Gray pursues the deep interest in the nonhuman world that makes his critique of humanism so sharp in fang and claw."
　　　　　　　　　　—Oliver Hall, *Dangerous Minds*

ALSO BY JOHN GRAY

Straw Dogs: Thoughts on Humans and Other Animals

Gray's Anatomy: Selected Writings

Black Mass: Apocalyptic Religion and the Death of Utopia

*The Immortalization Commission: Science and
the Strange Quest to Cheat Death*

The Silence of Animals: On Progress and Other Modern Myths

The Soul of the Marionette: A Short Inquiry into Human Freedom

Seven Types of Atheism

FELINE PHILOSOPHY

———

CATS AND
THE MEANING
OF LIFE

———

JOHN GRAY

PICADOR

FARRAR, STRAUS AND GIROUX   NEW YORK

Picador
120 Broadway, New York 10271

The Library of Congress has cataloged the Farrar, Straus and Giroux
hardcover edition as follows:
Names: Gray, John, 1948– author.
Title: Feline philosophy : cats and the meaning of life / John Gray.
Description: First American edition. | New York : Farrar, Straus and Giroux, 2020. |
    Includes bibliosgraphical references.
Identifiers: LCCN 2020027810 | ISBN 9780374154110
Subjects: LCSH: Animals (Philosophy) | Cats—Miscellanea. | Life.
Classification: LCC B105.A55 G73 2020 | DDC 128—c23
LC record available at https://lccn.loc.gov/2020027810

Paperback ISBN: 978-1-250-80025-1

Our books may be purchased in bulk for promotional, educational, or
business use. Please contact your local bookseller or the Macmillan Corporate
and Premium Sales Department at 1-800-221-7945, extension 5442,
or by email at MacmillanSpecialMarkets@macmillan.com.

For book club information, please visit facebook.com/picadorbookclub or
email marketing@picadorusa.com.

picadorusa.com • instagram.com/picador
twitter.com/picadorusa • facebook.com/picadorusa

7   9   10   8   6

Contents

1 Cats and Philosophy   1

    A cat-loving anti-philosopher:
    Michel de Montaigne   6

    Mèo's journey   9

    How cats domesticated humans   16

2 Why Cats Do Not Struggle to Be Happy   25

    When philosophers talk of happiness   26

    Pascal on diversion   31

    Hodge and the Fall   38

3 Feline Ethics   45

    Morality, a very peculiar practice   45

    Spinoza on living according to your nature   47

    Selfless egoism   58

4 Human vs Feline Love   67

    Saha's triumph   67

    Ming's biggest prey   70

    Loving Lily   75

    Gattino vanishes   79

5 Time, Death and the Feline Soul   89

    Muri's farewell   89

    Civilization as death-denial   93

    Cats as gods   99

CONTENTS

6   Cats and the Meaning of Life   105

    Cat nature, human nature   106

    Ten feline hints on how to live well   108

    Mèo on the window ledge   111

Acknowledgements   113

Notes   115

I

Cats and Philosophy

A philosopher once assured me he had persuaded his cat to become a vegan. Believing he was joking, I asked how he had achieved this feat. Had he supplied the cat with mouse-flavoured vegan titbits? Had he introduced his cat to other cats, already practising vegans, as feline role models? Or had he argued with the cat and persuaded it that eating meat is wrong? My interlocutor was not amused. I realized he actually believed the cat had opted for a meat-free diet. So I ended our exchange with a question: did the cat go out? It did, he told me. That solved the mystery. Plainly, the cat was feeding itself by visiting other homes and hunting. If it brought any carcasses home – a practice to which ethically undeveloped cats are sadly all too prone – the virtuous philosopher had managed not to notice them.

It is not hard to imagine how the cat on the receiving end of this experiment in moral education must have viewed its human teacher. Perplexity at the philosopher's behaviour would soon have been followed by indifference. Seldom doing anything unless it serves a definite purpose or produces immediate enjoyment, cats are arch-realists. Faced with human folly, they simply walk away.

The philosopher who believed he had persuaded his cat to adopt a meat-free diet only showed how silly philosophers can be. Rather than trying to teach his cat, he would have been wiser if he had tried learning from it. Humans cannot become cats. Yet if they set aside any notion of being superior beings, they may come to understand how cats can thrive without anxiously inquiring how to live.

Cats have no need of philosophy. Obeying their nature, they are content with the life it gives them. In humans, on the other hand, discontent with their nature seems to be natural. With predictably tragic and farcical results, the human animal never ceases striving to be something that it is not. Cats make no such effort. Much of human life is a struggle for happiness. Among cats, on the other hand, happiness is the state to which they default when practical threats to their well-being are removed. That may be the chief reason many of us love cats. They possess as their birthright a felicity humans regularly fail to attain.

The source of philosophy is anxiety, and cats do not suffer from anxiety unless they are threatened or find themselves in a strange place. For humans, the world itself is a threatening and strange place. Religions are attempts to make an inhuman universe humanly habitable. Philosophers have often dismissed these faiths as being far beneath their own metaphysical speculations, but religion and philosophy serve the same need.[1] Both try to fend off the abiding disquiet that goes with being human.

Simple-minded folk will say the reason cats do not practise philosophy is that they lack the capacity for abstract thought. But one can imagine a feline species that had this ability while still retaining the ease with which they inhabit the world. If these cats turned to philosophy, it would be as an amusing branch of fantastic fiction. Rather than looking to it as a remedy for anxiety, these feline philosophers would engage in it as a kind of play.

Instead of being a sign of their inferiority, the lack of

abstract thinking among cats is a mark of their freedom of mind. Thinking in generalities slides easily into a superstitious faith in language. Much of the history of philosophy consists of the worship of linguistic fictions. Relying on what they can touch, smell and see, cats are not ruled by words.

Philosophy testifies to the frailty of the human mind. Humans philosophize for the same reason they pray. They know the meaning they have fashioned in their lives is fragile and live in dread of its breaking down. Death is the ultimate breakdown in meaning, since it marks the end of any story they have told themselves. So they imagine passing on to a life beyond the body in a world out of time, and the human story continuing in this other realm.

Throughout much of its history, philosophy has been a search for truths that are proof against mortality. Plato's doctrine of forms – unchanging ideas that exist in an eternal realm – was a mystical vision in which human values were secured against death. Thinking nothing of death – while seeming to know well enough when it is time to die – cats have no need of these figments. If they could understand it, philosophy would have nothing to teach them.

A few philosophers have recognized that something can be learned from cats. The nineteenth-century German philosopher Arthur Schopenhauer (born in 1788) is famous for his love of poodles, a succession of which he kept throughout his later years, calling all of them by the same names – Atma and Butz. He also had at least one feline companion. When he died of heart failure in 1860, he was found at home on his couch beside an unnamed cat.

Schopenhauer used his pets to support his theory that selfhood is an illusion. Humans cannot help thinking of cats as separate individuals like themselves; but this is an error, he believed, since both are instances of a Platonic form, an archetype that recurs in many different instances. Ultimately each of these seeming individuals is an ephemeral embodiment of something more fundamental – the undying will to live, which, according to Schopenhauer, is the only thing that really exists.

He spelt out his theory in *The World as Will and Representation*:

> I know quite well that anyone would regard me as mad if I seriously assured him that the cat, playing just now in the yard, is still the same one that did the same jumps and tricks there three hundred years ago; but I also know that it is much more absurd to believe that the cat of today is through and through and fundamentally an entirely different one from that cat of three hundred years ago . . . For in a certain sense it is of course true that in the individual we always have before us a different being . . . But in another it is not true, namely in the sense in which reality belongs only to the permanent forms of things, to the Ideas, and which was so clearly evident to Plato that it became his fundamental thought.[2]

Schopenhauer's view of cats as fleeting shadows of an Eternal Feline has a certain charm. Yet when I think of the cats I have known it is not their common features that come first to mind but their differences from one another. Some cats are meditative and restful, others intensely playful; some cautious, others recklessly adventurous; some quiet and peaceable, others vocal and highly assertive. Each has its own tastes, habits and individuality.

Cats have a nature that distinguishes them from other creatures – not least ourselves. The nature of cats, and what we can learn from it, is the subject of this book. But no one who has lived with cats can view them as interchangeable instances of a single type. Every one of them is singularly itself, and more of an individual than many human beings.

Still, Schopenhauer was more humane in his view of animals than other leading philosophers. According to some reports, René Descartes (1596–1650) hurled a cat out of a window in order to demonstrate the absence of conscious awareness in non-human animals; its terrified screams were mechanical reactions, he concluded. Descartes also performed experiments on dogs, whipping one while a violin was being played in order to see whether the sound of a violin would later frighten the animal, which it did.

Descartes coined the expression, 'I think, therefore I am.' The implication was that human beings are essentially minds and only accidentally physical organisms. He wanted his philosophy to be based on methodical doubt. It did not occur to him to doubt the Christian orthodoxy that denied animals souls, which he renewed in his rationalist philosophy. Descartes believed his experiments proved non-human animals were insensate machines: what they actually showed is that humans can be more unthinking than any other animal.

Conscious awareness can spring up in many living things. If one strand in natural selection led to humans, another led to the octopus. There was nothing preordained in either case. Evolution is not moving towards increasingly self-aware forms of life. Appearing by chance, consciousness comes and goes in the organisms that possess it.[3] Twenty-first-century transhumanists think of evolution as leading to a fully self-aware cosmic mind. Such views have precedents in nineteenth-century theosophy, occultism and spiritualism.[4] None of them has any basis in Darwin's theory. The self-awareness of humans may be a one-off fluke.[5]

This may seem a bleak conclusion. But why should self-awareness be the most important value? Consciousness has been overrated. A world of light and shadow, which inter-mittently produces creatures that are partially self-aware, is more interesting and worth living in than one that basks in the unwavering radiance of its own reflection.

When turned in on itself, consciousness stands in the way of a good life. Self-consciousness has divided the human mind in an unceasing attempt to force painful experiences into a part that is sealed off from awareness. Suppressed pain festers in questions about the meaning of life. In contrast, the feline mind is one and undivided. Pain is suffered and forgotten, and the joy of life returns. Cats do not need to examine their lives, because they do not doubt that life is worth living. Human self-consciousness has produced the perpetual unrest that philosophy has vainly tried to cure.

A CAT-LOVING ANTI-PHILOSOPHER: MICHEL DE MONTAIGNE

A better understanding of cats, and of the limits of philoso-phy, was shown by Michel de Montaigne (1533–92), who wrote: 'When I play with my cat, how do I know that she is not passing time with me rather than I with her?'[6]

Montaigne is often described as one of the founders of modern humanism – a current of thought that aims to leave any idea of God behind. In fact he was as sceptical of humankind as he was about God. 'Man is the most blighted and frail of all creatures,' he wrote 'and, moreover, the most given to pride.' Scanning through past philosophies, he found none that could replace the knowledge of how to live that animals possess by nature. 'They may reckon us to be brute beasts for the same reason that we reckon them to be so.'[7] Other animals were superior to humans in possessing an innate understanding of how to live. Here Montaigne departed from Christian belief and the main traditions of western philosophy.

Being a sceptic in Montaigne's day was a risky business. Like other European countries France was wracked by wars of religion. Montaigne was drawn into them when he followed his father to become mayor of Bordeaux, and continued to act as a mediator between warring Catholics and Protestants after he retreated from the world to his study in 1570. Montaigne's family lineage included Marranos – Iberian Jews, who under persecution from the Inquisition were forced to convert to Christianity – and when he wrote in support of the Church he may have been safeguarding himself against the repression they suffered. At the same time he belongs in a tradition of thinkers who were open to faith because they doubted reason.

Ancient Greek scepticism was rediscovered in Europe in the fifteenth century. Montaigne was influenced by its most radical strand, Pyrrhonism, named after Pyrrho of Elis (c.360–c.270 BC), who travelled with the army of Alexander the Great to India, where he is reputed to have studied with the gymnosophists ('naked sages') or yogis. It may have been from these sages that Pyrrho imported the idea that the aim of philosophy was *ataraxia*, a term signifying a state of tranquillity, which he may have been the first to use. Suspending belief and disbelief, the sceptical philosopher could be safe from inner disturbance.

Montaigne learned much from Pyrrhonism. He had the beams of the tower to which he retreated in later life decorated with quotations from Pyrrho's follower, the physician-philosopher Sextus Empiricus (AD c.160–c.210), author of *Outlines of Pyrrhonism*, where the sceptical outlook was summarized:

The causal principle of scepticism we say is the hope of becoming tranquil. Men of talent, troubled by the anomaly in things and puzzled as to which of them they should rather assent to, came to investigate what in things is true and what false, thinking that by deciding these issues they would become tranquil.[8]

But Montaigne questioned whether philosophy, even of a Pyrrhonian kind, could deliver the human mind from turmoil. In many of his essays – a term Montaigne invented, coming from the French *essais*, meaning 'trials' or 'attempts' – he used Pyrrhonism in support of faith.

According to Pyrrho, nothing can be known. As Montaigne put it, 'There is a plague on Man: his opinion that he knows something.'[9] Pyrrho's disciples were taught to live by relying on nature rather than any argument or principle. But if reason is powerless, why not accept the mysteries of religion?

All of the three main schools of philosophy in the ancient European world – Stoicism, Epicureanism and Scepticism – had a state of tranquillity as their goal. Philosophy was a calmative, which if taken regularly would produce *ataraxia*. The end of philosophizing was peace. Montaigne had no such hopes: 'All the philosophers of all the sects are in general accord over one thing: that the sovereign good consists in peace of mind and body. But where are we to find it? . . . For our portion we have been allotted wind and smoke.'[10]

More sceptical than the most radical Pyrrhonist, Montaigne did not believe any philosophizing could cure human disquiet. Philosophy was useful chiefly in curing people of philosophy. Like Ludwig Wittgenstein (1889–1951), he recognized that ordinary language is littered with residues of past metaphysical systems.[11] By uncovering these traces and recognizing that the realities they describe are actually fictions, we could think more flexibly. Small doses of such a homoeopathic remedy against philosophy – an anti-philosophy, one might say – might bring us closer to other animals. Then we might be able to learn something from creatures that philosophers have dismissed as our inferiors.

An anti-philosophy of this kind would begin not with arguments, but with a story.

MÈO'S JOURNEY

The cat entered the room as a silhouette, a small black shape framed against the harsh light coming in from the doorway. Outside a war was blasting away. This was the Vietnamese city of Hué in February 1968 at the start of the Tet Offensive, the North Vietnamese campaign against American forces and their South Vietnamese allies that would lead to America's departure from the country five years later. In *The Cat from Hué*, one of the great accounts of the human experience of war, the CBS television journalist John (Jack) Laurence described the city:

> Hué was war fighting at its most ferocious. In this case, it was an urban brawl between two armed and largely adolescent tribes, both new to the territory and intent on taking it, a street fight of fast action and merciless bloodletting. There were no rules. Lives were taken without thought – snuffed, wasted, zapped . . . At the end, the more violent powerful gang drove the other off and claimed what was left. The losers withdrew with their casualties and lived to fight another day. The winners got the ruins. So it was in Hué.[12]

As it edged into the room the dark form could be seen to be a kitten, around eight weeks old, slight enough to fit into Laurence's hand. Skinny and dirty, its fur matted and greasy, the cat sniffed the air, catching the smell of the food the American journalist was eating from an army-issued can. The journalist tried talking in Vietnamese to the kitten, which looked back at him as if he were deranged. He offered it some of the food, which it approached cautiously but did not touch. Leaving some behind, the American left and came back the next day. The kitten appeared in the doorway, surveyed the room and walked towards him, sniffing his hand as he held out his fingers. All he had left to eat was a can marked 'beef slices', which he opened and offered on his fingers. The kitten ate ravenously, swallowing the slices of cooked meat without chewing them. Then the American soaked a towel in water from a canteen, and held the little cat by the shoulders, digging the dirt and fleas out of its ears, washing the filth from its mouth and rubbing its chin and whiskers clean. The kitten did not resist, and when the cleaning was done it licked the fur on its foreleg and washed its face. That done, it approached closer to the American and licked the back of his hand.

A jeep arrived, and Jack realized he was on his way home. He put the kitten in his pocket, and began a companionship that took them out of Hué by helicopter to Danang, where the kitten – now called Mèo, pronounced *may-oh* – lived in the press compound, eating four or five hearty meals a day. Along the way, Mèo scratched through the material of Jack's jacket and nearly escaped, explored the cockpit and climbed up the pilot's straps. They went on to Saigon, and this time Mèo travelled in a cardboard box containing his blanket and toys, unable to roam the plane and howling all the way. They stayed in a hotel together where Mèo had a much-resisted bath. His seemingly black fur proved to be an involuntary disguise, from which he emerged as a crossbred, red point Siamese with brilliant blue eyes.

In the hotel, Mèo was fed regularly – four meals a day of leftover fish heads and rice from the kitchen – though he made forays into other rooms in search of more to eat. He would jump onto the window ledge of the hotel room and lie there for hours, fully alert but almost motionless, his eyes following the movements of the people, lights and vehicles below. The American journalists involved in the war learned to endure it by getting high, drinking together and passing out, only to be woken by nightmares. At times they returned home for time off, but the war went with them and still disturbed their sleep. For his part, Mèo 'appeared to understand what was going on better than any of us from the outside . . . And that gave him his freedom, even in captivity. As he sat by the open window . . . enveloped in a fine haze of cigarette smoke, Mèo's eyes were as deep and blue and numinous as the South China Sea.'[13]

He slept in a bunker he had made for himself, a cardboard shipping carton into which he chewed a hole – a task that took a week – just big enough for him to squeeze through. He dominated the dozen or so feral cats in the hotel grounds, which learned to avoid him, and used the garden and rooms as hunting territory where he caught and ate lizards, pigeons, insects, snakes and possibly even a peacock, which mysteriously disappeared. His teeth now as sharp as daggers, he was 'the small white hunter, a natural-born killer, an ambush waiting to happen'.[14] Other than the Vietnamese hotel staff who came to feed him, he was hostile to anyone who came into the room, especially if they were American. 'He appeared to have a grudge against humanity . . . Withdrawn and isolated, hostile toward all but the Vietnamese, he was a wild malevolent animal, a singularly deep and inscrutable cat.'[15]

He had no fear and was never caught when entering other rooms. Jack came to see him as the reincarnation of Sun Tzu, the author of *The Art of War*, 'Intelligent, daring, cunning, ferocious . . . a VietCong version of the Chinese warrior-philosopher in the body of a cat . . . As a half-grown cat, he was tough, independent, irascible. Soldierly and serene. A Zen warrior in white fur . . . recklessness was part of his charm . . . Walking along the outside ledge of the hotel, attacking larger animals, setting traps with wicked guile, he risked his life with the casual abandon of those who think they're invincible . . . He was never nervous and never wasted energy. His moves were fluid, unfathomable.'[16]

When he adopted Mèo, Jack felt he was affirming life in a situation where it was being destroyed on an enormous scale:

> By providing food and shelter for the cat, I was affirming a life, however small and insignificant, in the midst of the slaughter. It wasn't conscious. Being young, I didn't dwell on my motives for doing things. It seemed right at the time. Though Mèo and I regarded each other as enemies, in a curious way we had come to depend on each other, just by being around, a kind of security in adversity. When I came back to the room after a trip to the field and heard him moving in his bunker or drinking water out of the tap in the bathroom or knocking something off the desk, it felt like coming home, belonging, feeling safe. Unprovoked attacks on me became less frequent, less ferocious, more of a ritual. Making it through Hué together must have formed a bond. Taking care of him gave me one small purpose other than reporting misery all the time.[17]

When he returned home in May 1968, Jack had Mèo follow him in the cargo hold of a later flight. If Mèo had stayed in Saigon he would most likely have joined countless other animal casualties of the war – the unknown numbers of dogs, monkeys, water buffaloes, elephants, tigers and other cats that were killed in the course of the conflict. If the Vietcong mounted another offensive, food would be short. Mèo could well end up in a cooking pot. So Jack took him to the Saigon Zoo, nearly empty as some of the animals had starved to death during the last offensive and few visitors came any more, where Mèo had the injections required in order to be certificated as safe to travel. A few days later he made the thirty-six-hour journey, screeching and scratching, to New York. When Jack picked him up and released him in his car, he jumped on the dashboard and clambered on Jack's shoulder, sniffing everywhere and observing the passing traffic. Arriving at Jack's mother's house in Connecticut, he consumed a can of American tuna.

Mèo settled well in his new home, scaring off other cats, hunting, and attacking unfamiliar adults while playing harmlessly with the local children. The household in turn adapted to Mèo. He was terrified of the sound of the vacuum cleaner, which may have reminded him of a tank or a plane, so the cleaner was not used when he was nearby. After Mèo pounced on her, the housekeeper quit. When he disappeared, Jack's mother searched for several days until he was discovered in a box in the garage, having somehow found his way there after a bad traffic accident.

The vet was not hopeful. Mèo's shoulder was shattered, and he needed an expensive operation in the animal hospital. But after six weeks in the hospital he returned to Jack's mother's house, where he inspected his favourite spots and resumed his life of tree-climbing, sleeping in the sun and hunting. His recovery continued until a bout of pneumonia signalled by violent sneezing and a loss of interest in food sent him back to the hospital for another three weeks. Forbidden treats were smuggled in and the staff made a fuss of him. This time he returned to full health, though for the rest of his life he had a habit of sneezing.

Having recovered, Mèo left Connecticut to join Jack in a one-bedroom apartment in an old brownstone house in Manhattan, where Jack lived with his partner Joy. In 1970 Jack returned to Vietnam for a month and Mèo seemed to miss him. When he came back, Mèo paid no attention to him. He sniffed Jack's luggage closely, as if it reminded him of something. Jack gave him a toy from Saigon, but he ignored it, went into his bunker and spent the rest of the afternoon there. In the evening, though, Joy told Jack, Mèo climbed on the bed, sat near Jack's head and spent hours looking at his face while he slept.

Back in America, Jack recalled his time in Vietnam with excitement and horror. He dulled his nightmares with drugs and alcohol. By the start of the 1970s New York was becoming dangerous, and at times it seemed he was back in a combat zone. When a job became vacant in London, he applied for it. Mèo followed Jack and Joy to London, where the couple had two daughters. He was forced to spend six months in quarantine, a trial he never forgot or forgave, despite Jack and Joy visiting him regularly. When he came to live with them again, he was wilder than before, tearing about in their London flat. When sleeping he would sometimes stiffen and shiver, 'as if . . . wrestling with ghosts'.[18]

After a time, Mèo settled down to a life of comfort and security with Jack and Joy and their two small children. One of Jack's daughters, Jessica, gave Mèo treats between meals, and Mèo slept with her at night. By then treating Jack as an old friend, Mèo would lick drops of whisky from his fingers late at night and retire to sleep at the same time. Mèo lived on until 1983, when a second bout of pneumonia proved fatal. Jack thought he would have preferred a warmer climate. It was the English weather that did for him.

He remembered Mèo,

> alone in the night, wandering through the far end of the apartment, making a cry that was unlike any other sound he made, unlike a sound I had heard any animal make. It seemed to be the call of an animal taken out of the wild, or out of its home, or away from its family. It was more of a wail, a long powerful howl, not a scream or a meow or an ordinary cat cry, but a call from the deepest part of his soul, the wail of the forest. The only time Mèo cried like that was when the home was quiet, usually when everyone was asleep, when he thought he was alone. It was a call for no one but himself.[19]

As Mèo made his intrepid journey through the world, humankind continued its random walk. Not long after he left Vietnam, the ancient and beautiful city of Hué was razed to the ground, with an unnamed American major commenting to a journalist, 'It became necessary to destroy the town to save it.' In what came to be known as 'the Hué massacre', North Vietnamese forces killed thousands of the inhabitants (the exact number is unknown). The Americans used the defoliant Agent Orange, destroying forests – the habitat of countless animal species – and producing genetic defects in humans. Over 58,000 US soldiers died in the conflict. Around 2 million Vietnamese civilians were killed. Unnumbered others were injured, disabled and traumatized.

Throughout the smoke and wind of history, Mèo lived his fierce, joyous life. Torn from his home by human madness, he flourished wherever he found himself.

Jack wrote:

I think we had come to respect each other's skills as survivors. There was no doubt that his limited number of lives allotted had been used up long ago, so that every new day he lived was a bonus. Also, he seemed wise. He knew. We had become friends. Our long, angry, loving relationship had come to symbolize in some way the bond between our countries, drenched in each other's blood, locked in an unbreakable embrace of life, suffering and death.[20]

HOW CATS DOMESTICATED HUMANS

At no point were cats domesticated by humans. One particular type of cat – *Felis silvestris*, a sturdy little tabby – has spread world-wide by learning to live with humans. House cats today are offshoots of a particular branch of this species, *Felis silvestris lybica*, which began to cohabit with humans some 12,000 years ago in parts of the Near East that now form part of Turkey, Iraq and Israel. By invading villages in these areas, these cats were able to turn the human move to a more sedentary life to their advantage. Preying on rodents and other animals attracted by stored seeds and grains and snatching waste meat left behind after slaughtered animals had been eaten, they turned human settlements into reliable food sources.

Recent evidence points to a similar process taking place independently in China around five millennia ago, when a central Asian variety of *Felis silvestris* pursued a similar strategy. Having entered into close proximity with humans, it was not long before cats were accepted as being useful to them. Employing cats for pest control on farms and sailing vessels became common. Whether as rat-catchers, stowaways or accidental travellers, cats spread on ships to parts of the world where they had not lived before. In many countries today, they outnumber dogs and any other animal species as cohabitants of human households.[21]

Cats initiated this process of domestication, and on their own terms. Unlike other species that foraged in early human settlements, they have continued to live in close quarters with humans ever since without their wild nature changing greatly. The genome of house cats differs in only a small number of ways from that of its wild kin. Their legs are somewhat shorter and their coats more variously coloured. Even so, as Abigail Tucker has noted, 'Cats have changed so little physically during their time among people that even today experts often can't tell house tabbies from wild cats. This greatly complicates the study of cat domestication. It's all but impossible to pinpoint the cats' transition into human life by examining ancient fossils, which hardly change even into modernity.'[22]

Unless they are kept indoors, the behaviour of house cats is not much different from that of wild cats. Though the cat may regard more than one house as home, the house is the base where it feeds, sleeps and gives birth. There are clear territorial boundaries, larger for male cats than for females, which will be defended against other cats when necessary. The brains of house cats have diminished in size compared with their wild counterparts, but that does not make house cats less intelligent or adaptable. Since it is the part of the brain that includes the fight-or-flight response that has shrunk, house cats have become able to tolerate situations that would be stressful in the wild, such as encountering humans and unrelated cats.

One reason cats were accepted by humans was their usefulness in reducing rodent populations. Cats eat rodents, and thousands of years ago were already eating mice that had eaten grain from human food stores. Yet in many environments cats and rodents are not natural enemies, and when they interact they often share a common resource such as household garbage. Cats are not very efficient as a means of pest control. House mice may have co-evolved with house cats, and learned to coexist with them. There are photographs of cats and mice together, only inches apart, in which the cats show no interest in the mice at all.[23]

A more fundamental reason why humans accepted cats in their homes is that cats taught humans to love them. This is the true basis of feline domestication. So beguiling are they that cats have often been seen as coming from beyond this world. Humans need something other than the human world, or else they go mad. Animism – the oldest and most universal religion – met this need by recognizing non-human animals as our spiritual equals, even our superiors. Worshipping these other creatures, our ancestors were able to interact with a life beyond their own.

Since their domestication of humans, cats have not needed to rely on hunting for their food. Yet cats remain hunters by nature, and when sustenance is not available from humans they soon return to a hunting life. As Elizabeth Marshall Thomas writes in *The Tribe of Tiger: Cats and Their Culture*, 'The story of cats is a story of meat.'[24] Big or small, cats are hyper-carnivores: in the wild, they only eat meat. That is why big cats are so endangered at the present time.

The rise of human numbers means expanding human settlements and shrinking open spaces. Cats are highly adaptable creatures, thriving in jungles, deserts and mountains as well as the open savannah. In evolutionary terms they have been extremely successful. Yet they are also extremely vulnerable. When their habitats and sources of food cease to be available, they are forced into conflicts with humans they are bound to lose.

Hunting and killing their food is instinctive in cats, and when kittens play it is hunting they are playing at. Cats need meat to live. They can digest vital fatty acids only when these are found in the flesh of other animals. The meat-free life of the moralizing philosopher would be death to cats.

How cats hunt tells us a good deal about them. Apart from lions, which hunt in packs, cats hunt alone, stalking and ambushing their prey, often at night. As ambush predators, cats have evolved for agility, jumping and pouncing in the pursuit of smaller prey. Wolves – the evolutionary ancestors of dogs – hunt for larger prey in groups held together by relationships of dominance and submission. Male and female wolves may mate for life, and both take care of offspring. None of these features of wolf behaviour is found in cats. The way cats relate to one another follows from their nature as solitary hunters.

It is not that cats are always alone. How could they be? They come together to mate, they are born in families and where there are reliable food sources they may form colonies. When several cats live in the same space a dominant cat may emerge. Cats may compete ferociously for territory and mates. But there are none of the settled hierarchies that shape interactions among humans and their close evolutionary kin. Unlike chimps and gorillas, cats do not produce alpha specimens or leaders. Where necessary, they will cooperate in order to satisfy their wants, but they do not merge themselves into any social group. There are no feline packs or herds, flocks or congregations.

That cats acknowledge no leaders may be one reason they do not submit to humans. They neither obey nor revere the human beings with which so many of them now cohabit.

Even as they rely on us, they remain independent of us. If they show affection for us, it is not just cupboard love. If they do not enjoy our company, they leave. If they stay, it is because they want to be with us. This too is a reason why many of us cherish them.

Not everyone loves cats. In recent times they have been demonized as 'an environmental contaminant ... like DDT', [25] which spread diseases such as rabies, parasitic toxoplasmosis and the pathogens responsible for the Black Death. Bird droppings pose a greater risk to human health, but one of the commonest accusations against cats is that they kill so many birds. The case against them is that they disrupt the balance of nature. Yet it is hard to explain hostility to cats in terms of any risks they may pose to the environment.

The danger of disease can be countered by programmes such as trap-neuter-return (TNR), widely implemented in the US, in which cats living outdoors are brought to clinics for vaccination and spaying and then released. The risk to birds can be diminished by bells and similar devices. More to the point, it is strange to single out one branch of a non-human species as a destroyer of ecological diversity when the major culprit in this regard is the human animal itself. With their superlative efficiency as hunters, cats may have altered the ecosystem in parts of the world. But it is humans that are driving the planetary mass extinction that is currently underway.

Hostility to cats is not new. In early modern France it inspired a popular cult. Cats had long been linked with the devil and the occult. Religious festivals were often rounded off by burning a cat in a bonfire or throwing one off a roof. Sometimes, in a demonstration of human creativity, cats were hung over a fire and roasted alive. In Paris it was the custom to burn a basket, barrel or sack of live cats hung from a tall mast. Cats were buried alive under the floorboards when houses were built, a practice believed to confer good fortune on those who lived there.[26]

On New Year's Day 1638, in Ely Cathedral, a cat was roasted alive on a spit in the presence of a large and boisterous crowd. A few years later Parliamentary troops, fighting against Royalist forces in the English Civil War, used hounds to hunt cats up and down Lichfield Cathedral. During pope-burning processions in the reign of Charles II, the effigies were stuffed with live cats so that their screams would add dramatic effect. At rural fairs a popular sport was shooting cats suspended in baskets.[27]

In some French cities, cat-chasers put on a livelier show by setting fire to them and pursuing them as they were burning through the streets. In other entertainments, cats would be passed around so that their fur could be torn off. In Germany the howls of cats tortured during similar festivals was called *Katzenmusik*. Many carnivals concluded with a mock trial in which cats were bludgeoned half to death and then hanged, a spectacle that evoked riotous laughter. Often cats were mutilated or killed as embodiments of forbidden sexual desire. From St Paul onwards, Christians viewed sex as a disruptive and even demonic force. The freedom of cats from human moralities may have become linked in the medieval mind with the rebellion of women and others against religious prohibitions on sex. Against the background of this kind of theism it was almost inevitable that cats should be seen as embodiments of evil. Throughout much of Europe they were identified as agents of witchcraft and tormented and burned along with or instead of witches.[28]

The practice of torturing cats did not end with the witchcraft craze. The nineteenth-century Italian neurologist Paolo Mantegazza (1831–1910), professor at the Istituto di Studi

Superiori in Florence, founder of the Italian Anthropological Society and later a progressive member of the Italian Senate, was an avowed Darwinian who believed humans had evolved into a racial hierarchy with 'Aryans' at the top and 'Negroids' at the bottom. The distinguished professor devised a machine he jovially entitled 'the tormentor'. Cats were 'quilted with long thin nails' so that any movement was agony, then flayed, lacerated, twisted and broken until death at last released them. The aim of the exercise was to study the physiology of pain. Like Descartes, who refused to abandon the theistic dogma that animals have no soul, the eminent neurologist believed that the torture of animals was justified by the pursuit of knowledge. Science perfected the cruelties of religion.[29]

At bottom, hatred of cats may be an expression of envy. Many human beings lead lives of muffled misery. Torturing other creatures is a relief, since it inflicts worse suffering on them. Tormenting cats is particularly satisfying, since they are so satisfied in themselves. Cat-hatred is very often the self-hatred of misery-sodden human beings redirected against creatures they know are not unhappy.

Whereas cats live by following their nature, humans live by suppressing theirs. That, paradoxically, is their nature. It is also the perennial charm of barbarism. For many human beings, civilization is a state of confinement. Ruled by fear, sexually starved and filled with rage they dare not express, such people cannot help being maddened by a creature that lives by affirming itself. Tormenting animals diverts them from the dismal squalor in which they creep through their days. The medieval carnivals in which cats were tortured and burned were festivals of the depressed.

Cats are disparaged for their apparent indifference to those that care for them. We give them food and shelter, yet they do not regard us as their owners or their masters, and they give us nothing back except their company. If we treat them with respect, they grow fond of us, but they will not miss us when we are gone. Lacking our support, they soon re-wild. Though they display little concern for the future, they seem set to outlast us. Having spread across the planet on the ships human beings used to expand their reach, cats look like being around long after humans and all their works have vanished without trace.

2

Why Cats Do Not Struggle to Be Happy

When people say their goal in life is to be happy they are telling you they are miserable. Thinking of happiness as a project, they look for fulfilment at some future time. The present slips by, and anxiety creeps in. They dread their progress to this future state being disrupted by events. So they turn to philosophy, and nowadays therapy, which offer relief from their unease.

Posing as a cure, philosophy is a symptom of the disorder it pretends to remedy. Other animals do not need to divert themselves from their condition. Whereas happiness in humans is an artificial state, for cats it is their natural condition. Unless they are confined within environments that are unnatural for them, cats are never bored. Boredom is fear of being alone with yourself. Cats are happy being themselves, while humans try to be happy by escaping themselves.

It is here that cats are most different from humans. As the founder of psychoanalysis Sigmund Freud understood, an uncanny sort of misery is normal for human beings. Freud never explained this condition, or believed psychoanalysis could cure it. Today there are numberless techniques that promise deliverance from it. These therapies may equip people to live less uncomfortably with other human beings. They cannot rid them of the unrest that goes with being human.

This is why so many humans love being with cats. Ailuro-philes are often accused of anthropomorphism – the practice of attributing human emotions to other animals that lack them. But cat-lovers do not love cats because they recognize themselves in them. They love cats because cats are so different from them.

Unlike dogs, cats have not become part-human. They interact with us and may in their own way come to love us, but they are other than us in the deepest levels of their being. Having entered the human world, they allow us to look beyond it. No longer trapped within our own thoughts, we can learn from them why our nervous pursuit of happiness is bound to fail.

WHEN PHILOSOPHERS TALK OF HAPPINESS

Philosophy has rarely been an open-ended inquiry. In medie-val times it was a servant of theology. Today it is the practice of elucidating the prejudices of middle-class academics. In its earliest forms it aimed to teach tranquillity.

Among the ancient philosophers, the Epicureans believed they could achieve happiness by curbing their desires. When someone is described as an epicurean today, we think of a person who delights in fine food and wine and the other pleasures of life. But the original Epicureans were ascetics who aimed to reduce their pleasures to a min-imum. They dined on a simple diet of bread, cheese and olives. They had no objection to sex as long as it was used medicinally, as a remedy against frustration, and not mixed with infatuation or what we would nowadays call romantic love, which would only disturb their peace of mind. For the same reason, they spurned any form of ambition or political engagement. Withdrawal into the tranquil seclusion of a well-appointed garden would secure them from pain and anxiety and enable them to achieve *ataraxia*.

Epicurus has some things in common with the Buddha. Both promise release from suffering through the abandonment of desire. But the Buddha is more realistic in acknowledging that this can be fully achieved only by stepping off the carousel of birth and death – in other words, by ceasing to exist as a distinct individual. Enlightened human beings may experience a state of bliss during their lifetime; but they can be liberated from suffering only when they are no longer going to be reborn.

If you accept the myth of reincarnation, this story may have some appeal. The Epicurean vision is harder to take seriously. For Epicurus and his disciples the universe is a chaos of atoms floating in a void. Gods may exist, but they are indifferent to us. The task of human beings is to remove the sources of suffering in themselves. So far, this is much like Buddhism. The difference is that Epicurus can promise release only from the sufferings that come from mistaken beliefs and excessive desires. Death may be greeted with acceptance, as in the case of Epicurus himself, who remained cheerful and continued to teach throughout his final illness. But it is not clear what Epicurus has to say to those who suffer from unending hunger and overwork, persecution or poverty.

You can enjoy Epicurean seclusion only if you live in a time and place that permits such luxury and you are lucky enough to be able to afford it. This has never been true for most human beings, nor will it ever be. Where such retreats have existed they have been shelters for the few, and overrun in wars and revolutions. A more fundamental limitation of Epicurean philosophy is the spiritual poverty of the life it recommends. It is a neurasthenic vision of happiness. As in a convalescent hospital, no noise is allowed. Only a restful stillness remains. But then life is stilled, and much of its joy is gone.

The Spanish-American philosopher George Santayana captured this poverty when discussing the Roman poet-philosopher Lucretius, who presented Epicurus' vision in his poem *On the Nature of Things*:

> Lucretius' notion . . . of what is positively worth while or attainable is very meagre: freedom from superstition, with so much natural science as may secure that freedom, friendship, and a few cheap and healthful animal pleasures. No love, no patriotism, no enterprise, no religion.[1]

Epicureans aimed to achieve tranquillity by paring away the goods of life to the point at which (these sages imagined) the few that remained could be enjoyed in all circumstances. Stoics approached the same end by a different route. By controlling their thoughts, they believed, they could accept anything that happened to them. The cosmos was governed by Logos, or reason. If you feel an event is disastrous, it is because you have not understood that it is part of the cosmic order. The way to tranquillity is to identify yourself with this order. Having done this, you can find fulfilment in playing your part in the scheme of things.

This Stoic philosophy found followers in many parts of society, from slaves to rulers. A picture of how it was used can be found in the *Meditations* of the emperor Marcus Aurelius (AD 121–180). A spiritual diary in which he exhorts himself to accept his place in the world and do his duty, the *Meditations* is infused with weariness of life. Marcus urges himself to consider:

How all things are vanishing swiftly, bodies themselves in the Universe and the memorials of them in Time; what is the character of all things of sense, and most of all those which attract by the bait of pleasure or terrify by the threat of pain or are shouted abroad by vanity, how cheap, contemptible, soiled, corruptible, and mortal: these are for the faculty of mind to consider. To consider too what kind of men those are whose judgements and voices confer honour and dishonour; what it is to die, and that if a man looks at it by itself and by the separating activity of thought strips off all the images associated with death, he will come to judge it to be nothing else but Nature's handiwork.[2]

This is not an affirmation of life but a pose of indifference to life. By composing in his mind a rational scheme of things in which he is a necessary part, Aurelius struggles to reconcile himself to misfortune and death. The emperor-philosopher believed that if he could find a rational order within himself he would be saved from anxiety and despair. For not only is the universe rational: what is rational is right and good. In this fictitious oneness Marcus hoped to find peace.

For Marcus, reason required a willed extinction of the will. The result is a funereal celebration of endurance and resignation. The philosopher-emperor dreams of becoming like an immovable statue in a hushed Roman mausoleum. But life wakes him from his dream, and he must weave a shroud of philosophy around himself all over again.

The Russian poet and essayist Joseph Brodsky wrote:

For the ancients, philosophy wasn't a by-product of life but the other way around . . . Perhaps we should momentarily dispense here with the very word 'philosophy', for Stoicism, its Roman version especially, shouldn't be characterized as love for knowledge. It was, rather, a lifelong experiment in endurance . . .[3]

Grimly performing his imperial duties – the part in life the cosmos had given him, he liked to believe – Marcus found satisfaction in contemplating his own sadness.

Stoics accepted that even the wisest sage could not endure the worst of life's pains. When this was the case, suicide was permissible. Marcus discouraged killing yourself if you had some public responsibility to perform, while allowing that you might end your life if circumstances made any kind of rational existence impossible for you.

The Stoic philosopher, statesman and dramatist Seneca went further, and believed suicide could be reasonable if you simply had had enough of living. Advising a young disciple, he asked:

> Have you anything worth waiting for? Your very pleasures, which cause you to tarry and hold you back, have already been exhausted by you. None of them is a novelty to you, and there is none that has not already become hateful because you are cloyed with it. You know the taste of wine and cordials. It makes no difference whether a hundred or a thousand measures pass through your bladder . . . It is with life as it is with a play, – it matters not how long the action is spun out, but how good the acting is. It makes no difference at what point you stop. Stop whenever you choose; only see to it that the closing period is well turned. Farewell.[4]

Seneca died by his own hand, though not by his own choice. Accused of complicity in a plot to kill the emperor Nero, he was ordered by Nero to commit suicide. According to the Roman historian Tacitus, Seneca obeyed and slit his veins. But the blood flowed slowly, so he took poison. This too failed, and he was placed by soldiers into a warm bath, where he finally suffocated.

As a way of living, *ataraxia* is an illusion. Epicureans try to simplify their life so as to reduce to a minimum the pleasures they can lose. But they cannot secure their tranquil garden against the turmoil of history. The Stoic sage insists that while we cannot control the events that happen to us, we can control how we think about them. But this is so only within a narrow margin. A fever, a tsetse fly or a traumatic experience can unsettle the mind at a crucial moment, or for ever. Disciples of Pyrrho try to establish inner equilibrium by a suspension of judgement. But sceptical doubt cannot banish the unrest that comes with being a human being.

Even if *ataraxia* could be achieved, it would be a listless way to live. Luckily, deathly calm is not in practice a state that humans can maintain for very long.

PASCAL ON DIVERSION

All these philosophies have a common failing. They imagine life can be ordered by human reason. Either the mind can devise a way of life that is secure from loss, or else it can control the emotions so that it can withstand any loss. In fact, neither how we live nor the emotions we feel can be controlled in this way. Our lives are shaped by chance and our emotions by the body. Much of human life – and much of philosophy – is an attempt to divert ourselves from this fact.

Diversion was a central theme in the writings of the seventeenth-century scientist, inventor, mathematician and religious thinker Blaise Pascal, who wrote:

Diversion. Being unable to cure death, wretchedness and ignorance, men have decided, in order to be happy, not to think about such things.[5]

Pascal explains:

Sometimes, when I set to thinking about the various activities of men, the dangers and troubles which they face at Court, or in war, giving rise to so many quarrels and passions, daring and often wicked enterprises and so on, I have often said that the sole cause of man's unhappiness is that he does not know how to stay quietly in his room. A man wealthy enough for life's needs would never leave home to go to sea or besiege some fortress if he knew how to stay at home and enjoy it . . .

But after closer thought, looking for the particular reasons for all our unhappiness . . . I found one very cogent reason in the natural unhappiness of our feeble mortal condition, so wretched that nothing can console us when we really think about it . . .

The only good thing for men therefore is to be diverted from thinking of what they are, either by some occupation which takes their mind off it, or by some novel and agreeable passion which keeps them busy, like gambling, hunting, some absorbing show, in short what is called diversion.[6]

Humans divert themselves by using their imagination:

Imagination. It is the dominant faculty in man, master of error and falsehood . . . I am not speaking of fools, but of the wisest men, amongst whom imagination is best entitled to persuade. Reason may object in vain, it cannot fix the price of things.

This arrogant force, which checks and dominates its enemy, reason, for the pleasure of showing off the power it has in every sphere, has established a second nature in man.

Imagination has its happy and unhappy men, its sick and well, its rich and poor; it makes us believe, doubt, deny reason; it deadens the senses, it arouses them; it has its fools and sages . . . Imagination decides everything: it creates beauty, justice and happiness, which is the world's supreme good.[7]

Montaigne also wrote on diversion. But whereas Pascal rejected it as an obstacle to salvation, Montaigne welcomed it as a natural remedy for suffering:

Once upon a time I was touched by a grief, powerful on account of my complexion and as justified as it was powerful. I might well have died from it if I had merely trusted to my own strength. I needed a mind-departing distraction to divert it; so by art and effort I made myself fall in love, helped in that by my youth. Love comforted me and took me away from the illness brought on by that loving-friendship. The same applies everywhere: some painful idea gets hold of me; I find it quicker to change it than to subdue it . . . If I cannot fight it, I flee it; and by my flight I made a diversion and use craft; by changing place, occupation and company I escape from it into the crowd of other pastimes and cogitations, in which it loses all track of me and cannot find me.

That is Nature's way when it grants us inconstancy . . .[8]

Montaigne's grief was occasioned by the death of his beloved friend, the French judge and political thinker Étienne de La Boétie (1530–63), about whom he wrote a celebrated essay.[9] He overcame the melancholy that ensued by following 'Nature's way'.

In regard to diversion, humans and cats are at opposite poles. Not having formed an image of themselves, cats do not need to divert themselves from the fact that they will some day cease to exist. As a consequence, they live without the fear of time passing too quickly or too slowly. When cats are not hunting or mating, eating or playing, they sleep. There is no inner anguish that forces them into constant activity. When sleeping, they may dream. But there is no reason to think they dream of being in any other world, and when they are not asleep they are wholly awake. A time may come when they know they are about to die, but they do not spend their lives dreading its arrival.

Montaigne and Pascal accept that philosophy cannot divert the human animal from its misery. They differ in their view of what this misery signifies. Whereas Montaigne thinks other animals are superior to humans in some ways, Pascal regards human misery as a sign that humans are superior to all other animals: 'Man's greatness comes from knowing he is wretched: a tree does not know it is wretched. Thus it is wretched to know that one is wretched, but there is greatness in knowing one is wretched . . . It is the wretchedness of a great lord, the wretchedness of a dispossessed king.'[10] Where Montaigne turns to nature, Pascal turns to God.

In the course of his short life, Pascal achieved some astonishing intellectual feats. Before dying in 1662 at the age of thirty-nine, he built some of the first calculating machines (leading to a twentieth-century programming language being named after him) and made significant advances in probability theory. He also designed the first urban mass-transit system using horse-drawn buses, which operated in Paris for a time, and an early version of the roulette wheel. He is rightly recognized as one of the founders of modern science. But Pascal's overriding concern was with religion.

On 23 November 1654 he had a mystical revelation, a direct experience of the God that before then had remained hidden from him, which became the pivotal event of his life. He recorded the experience on a scrap of paper, then a piece of parchment, which he carried with him for the rest of his life. Found sewn into his clothing after he died, the text can be read as one of the *Pensées*.[11]

Pascal's last years were painful. He had been attracted by Jansenism, a current within Catholicism condemned by the Pope, and when after lifelong sickliness he became fatally ill, he was subjected to pointless and painful treatments by incompetent doctors and denied the consolation of the sacrament until almost the end. After a long agony, he died on 19 August 1662, his last words being: 'May God never abandon me.'

Pascal devoted many of the *Pensées* to refuting Montaigne's scepticism. He aimed to show that the chronic anxiety from which human beings suffer is a sign that they do not belong in the natural world. It is wrong for humans to look up to other animals: 'It is dangerous to explain too clearly to man how like he is to the animals without pointing out his greatness. It is also dangerous to make too much of his greatness without his vileness.'[12] Worst of all is when humans revere animals as gods. 'Man is vile enough to bow down to beasts and even worship them.'[13]

For Pascal, human unease points beyond the world. For Montaigne, it comes from a flaw in the human animal. Here I side with Montaigne. Humans are self-divided creatures whose lives are mostly spent in displacement activity. The sorrows they have in common with their animal kin are multiplied by thought constantly doubling back on itself. It is this reflexive self-consciousness that engenders the special wretchedness of the human animal.

Like Montaigne, Pascal mocked the idea that reason could provide a remedy for the human condition. Yet he did think reason could play a part in bringing human beings to faith. Pascal's famous wager offers reasons why we should bet in favour of God's existence. We cannot help but gamble, one way or another: if we win, we gain infinite happiness; if there is no God, we lose a finite mortal life, so short as to amount to almost nothing.[14]

It is an argument that leaves something to be desired. Pascal assumes we already know which God to gamble on. But human beings have worshipped many gods, each demanding submission and obedience. If we gamble on one that does not exist, another may damn us. Again, are our brief lives worth so little? If they are all we have, they may be all the more precious to us.

One should not take Pascal's appeal to reason too seriously. Reason points to faith, he believed, but he knew reason could not keep anyone faithful. The basis of any lasting faith is ritual. Instead of thinking about religion, people should go to a church, temple or synagogue and kneel, worship and pray with others. Human beings are more like machines than they like to imagine:

> For we must make no mistake about ourselves: we are as much automaton as mind. As a result, demonstration is not the only instrument for convincing us ... Proofs only convince the mind; habit provides the strongest proofs, and those that are most believed. It inclines the automaton, which leads the mind unconsciously along with it. Who ever proved that it will dawn tomorrow, and that we shall die? And what is more widely believed? It is, then, habit that convinces us and makes so many Christians ... We must acquire an easier belief, which is that of habit.[15]

Belief is a habit of the body. If you want faith, act as if you have it already. The mind will soon follow. Practice will make your faith enduring.

The trouble is that Pascal's analysis also justifies diversion. He writes: 'The eternal silence of these infinite spaces fills me with dread.'[16] But surrendering to the way of the world – following sport, or throwing yourself into a new love affair – may be just as effective in staving off existential dread as practising a religion. Any pastime may do the trick.

Where Pascal is right is that diversion is a uniquely human trait. Some have believed tool-making marks us off from our animal kin. Others have claimed it is the transmission of knowledge or the use of language. But none of these is exclusively human. Beavers build homes for themselves, ravens use tools to seize food, apes form cultures using knowledge transmitted from previous generations. The howls of wolves and the songs of whales are the sounds of them speaking to one another. The need for diversion, though, is essentially human.

Diversion is a response to the defining feature of the human animal: the fear of death that comes with self-awareness. Along with some other animals, elephants may recognize something like death when it happens to other members of their species. But only humans know a day will come when they themselves will die. Our image of ourselves passing through time comes with the realization that we will soon pass away. Much of our lives are spent running from our own shadow.

The denial of death and the division of the human soul go together. Dreading anything that reminds them of their mortality, humans push much of their experience into an unconscious part of themselves. Life becomes a struggle to stay in the dark. Not needing such darkness within themselves, cats on the other hand are nocturnal creatures that live in the light of day.

HODGE AND THE FALL

Cats do not plan their lives; they live them as they come. Humans cannot help making their lives into a story. But since they cannot know how their life will end, life disrupts the story they try to tell of it. So they end up living as cats do, by chance.

Humans differ from other animals in making far-reaching provisions for the future. Through agriculture and industry, they are less dependent on the seasons and changing weather. As a result, their lives are longer than before. But the way they live continues to be fragile.

Many seem confident that the modern civilization that has developed over the past few hundred years will endure, even though climate change and global pandemic are creating a different and more dangerous world. No doubt humankind will somehow adapt. But the nature of these adjustments is unclear. Will the kinds of societies that exist today be renewed in different forms? Or will institutions from the past – feudalism or slavery, for example – be revived and maintained with new technologies? No one can tell. The future of human life on Earth is as unknown as what (if anything) comes after we die.

Some modern thinkers have imagined that society could be rebuilt in order that human beings could achieve the happiness to which they have come to believe they are entitled. One who doubted this vision was the eighteenth-century English novelist, biographer, lexicographer and conversationalist Samuel Johnson (1709–84).

Born the son of a bookseller, Johnson entered the world a sickly infant that was not expected to live long. Early in life he displayed facial and bodily tics, leading some biographers to conclude that he suffered from Tourette syndrome. Often in debt and always in need of money, he attended Pembroke College, Oxford, but learned little and left without a degree. In 1735 he married the widow of a close friend, Henry Porter, a wealthy Birmingham merchant. Elizabeth 'Tetty' Porter was more than twenty years older than Johnson, and their relationship was discouraged by her family and a surprise to Johnson's friends. But the marriage – 'a love-match', he said – seems to have been happy and lasted until Tetty died in 1752. She bankrolled Johnson in founding a school, but the venture failed and for the rest of his life he was a jobbing writer. He always referred to her with gratitude and affection.

Like Montaigne, Johnson was a devoted cat-lover. He would walk into town to buy oysters for Hodge, his black-furred feline companion, and valerian to ease the cat's pain when it was ill. Again like Montaigne, though more often and more severely, he suffered from attacks of melancholy.

Johnson mocked the belief that happiness could be achieved by thinking about the best path in life. As he wrote to his friend and biographer James Boswell:

> Life is not long, and too much of it must not pass in idle deliberation how it shall be spent; deliberation, which those who begin it by prudence, and continue it with subtilty, must, after long expence of thought, conclude by chance. To prefer one future mode of life to another, upon just reasons, requires faculties which it has not pleased our Creator to give us.[17]

Johnson spelt out this view in *The History of Rasselas, Prince of Abissinia* (1759). Originally entitled 'The Choice of Life', the book is a fable in which the son of the king of Abyssinia (modern-day Ethiopia) leaves 'the Happy Valley' in which he has lived and travels to other countries.

Until then, Rasselas had not known the evils of the world. Surrounded by peace and beauty, he inhabited a version of paradise. He became bored and discontented, and wanted to know why. Yet none of the people he met were happy, and the sages he encountered could not tell him how to be happy himself. Should he persist in his quest? His friend, the poet Imlac, who has accompanied him on his journey, explains to him why pursuing happiness is chasing an illusion:

> 'The causes of good and evil, answered Imlac, are so various and uncertain, so often entangled with each other, so diversified by various relations, and so much subject to accidents which cannot be foreseen, that he who would fix his condition upon incontestable reasons of preference, must live and die inquiring and deliberating . . .
>
> 'Very few, said the poet, live by choice. Every man is placed in his present condition by causes which acted without his foresight, and with which he did not always willingly co-operate . . .'[18]

The story ends with Rasselas giving up his search and returning to the Happy Valley.

Johnson's conviction that thought could not relieve unhappiness reflected his own experience. Throughout his life he was preoccupied with his health. He contracted scrofula, an infection of the lymph nodes often caused by tuberculosis, which produced swelling of the glands. He lost part of his eyesight. In an autobiographical note written in his fifties, he described himself as having been 'a poor diseased infant, nearly blind'. Short-sightedness was a serious disadvantage in one whose business was the written word. Working for nearly a decade, he authored a dictionary of the English language that is also a major work of literature. His housekeeper Mrs Thrale describes how, reading at night bent over a candle, 'the foretops of all his wigs were burned by the candle down to the very network'.

Though Johnson was a fervent Christian, his faith did not give him peace. Always prone to depression, he often feared losing his mind. He kept a chain and padlock with which he asked Mrs Thrale to fetter him when madness seemed imminent. Some have suggested he may have had masochistic tendencies, but it seems more likely that he dreaded disgrace, and possible confinement, if the disturbed state of his mind became public knowledge. His acquaintance, the poet Christopher Smart, had been confined in a lunatic asylum for seven years, his only regular companion his cat Jeoffry, to whom he dedicated his celebrated poem in praise of this member of 'the tribe of Tiger'.[19]

Johnson may have feared the madhouse as much as he did madness itself. But it is true that he lived with almost continuous fear of derangement, which he tried to keep at bay by obsessive-compulsive rituals. When walking the London streets, he touched every post with his cane; if he missed one, he had to start his walk again. When sitting, he would rock back and forth, sometimes whistling. Forever muttering and mumbling to himself, he was a mass of tics and twitches. Whether or not he suffered from Tourette syndrome, he was a profoundly unsettled human being.

Yet Johnson's unease was only an exaggerated version of a disquiet common to all human beings. Much of human life is a succession of tics. Careers and love affairs, travels and shifting philosophies are twitches in minds that cannot settle down. As Pascal put it, human beings do not know how to sit quietly in a room. Johnson knew he could never sit quietly anywhere, but he could not cure himself of his restlessness. Like other human beings, he was ruled by his imagination.

In Chapter 44 of *Rasselas*, Johnson analyses the imagination's dangerous power and concludes that it cannot be overcome by an act of will:

> Perhaps, if we speak with rigorous exactness, no human mind is in its right state. There is no man whose imagination does not sometimes predominate over his reason, who can regulate his attention wholly by his will, and whose ideas will come and go at his command. No man will be found in whose mind airy notions do not sometimes tyrannise, and force him to hope or fear beyond the limits of sober probability. All power of fancy over reason is a degree of insanity . . .[20]

For Johnson, the best diversion from the disordered imagination was company. He made friends with the human wreckage of London as easily as with literary high society. Having conversed with King George III, he was no less ready to talk with homeless beggars, whom he would take into his home. Not thought, but immersion in society allowed him to escape himself.

Is there any other animal that cannot bear its own company? Certainly not any cat. Cats spend much of their lives in contented solitude. Yet they can grow fond of their human companions, and may treat the sick unease in them that humans themselves cannot remedy. Johnson appreciated this power in his cat, and described him as 'a very fine cat, a very fine cat indeed'. Hodge gave him something human company could not supply: a glimpse of life before the Fall.

An eighteenth-century rendition of the Garden of Eden, the Happy Valley in *Rasselas* is a place no one can revisit. To be sure, the prince decides to return to the Valley. But the last chapter of the novella is entitled 'The conclusion, in which nothing is concluded', and it is plain that neither the prince nor the Happy Valley can be again what they were when he left.

You can be in paradise only when you do not know what it is like to be in paradise. As soon as you know, paradise is gone. No effort of thought can take you back, for thought – the conscious awareness of yourself as a mortal being – is the Fall. In the Garden of Eden, the primordial human pair are clothed in ignorance of themselves. When they come to self-awareness, they find they are naked. Thinking of yourself is the gift of the serpent that cannot be returned.

For the eighteenth-century wordsmith, paradise was a state of mind in which he was not tormented by his thoughts. But Johnson knew self-torment was congenital in him. The 'poor diseased infant' would never know health: all he could do was escape from himself. So he submerged himself in society. The consummate conversationalist that Boswell recorded, incandescent with wit, was Johnson in flight from his own thoughts. Yet Johnson needed something more than diversion, something only his feline companion could offer. When he heard 'a young gentleman of good family' had gone mad and was shooting cats, Johnson murmured, 'But Hodge shan't be shot; no, no, Hodge shall not be shot.' Hodge gave Johnson respite from thought, and so from being human.

3

Feline Ethics

MORALITY, A VERY PECULIAR PRACTICE

Cats are often described as being amoral. They obey no commandments and have no ideals. They show no signs of experiencing guilt or remorse, any more than they do of struggling to be better than they are. They do not exert themselves to improve the world, or agonize over what is the right thing to do. If they could understand it, the notion that how they live should be decided by any external standard would be laughable to them.

Many people claim to value morality more than anything else. In their view, nothing marks them off from their animal kin more than the sense of right and wrong. A good life is not just one that is worth living; it must also be *moral*. If a life does not satisfy the requirements of morality, it cannot have much worth – or possibly any worth at all. Morality deals with a special sort of value, incomparably more precious than any other. Pleasure may be valuable, as may beauty and life itself, but unless these goods are pursued morally they are worthless, or else positively bad. This is true for every human being, since the laws of morality are universal and categorical. Everyone must be moral before they can be anything else.

Those who think like this are convinced they know what morality dictates. There cannot be fundamental disagreement where right and wrong are concerned. After all, being moral is the supreme good. How could human beings differ on something so important? In fact, there are many divergent and opposing moralities. For some people today, justice is the core of morality. But justice is neither as unchanging nor as important to them as they imagine. As Pascal noted, 'Justice is as much a matter of fashion as charm is.'[1]

Morality has many charms. What could be more captivating than a vision of everlasting justice? Yet visions of justice are as immutable as styles in shoes. What morality demands shifts across the generations and may change more than once within a single human lifetime. Not so long ago morality required spreading civilization by extending imperial power. Today, morality condemns empire in all its forms. These judgements are irreconcilably opposed. But they provide the same satisfaction to those who pronounce them – a gratifying sense of virtue.

When people talk of morality they do not know what it is they are talking about. At the same time they are unshakably certain in what they say. This may seem paradoxical. But it is not, since what they are doing is expressing their emotions. Apart from any facts they may cite in support of them, there is nothing true or false in their judgements of value. That is why there cannot be agreement in morality. If value judgements merely express emotions, there is nothing to agree (or disagree) about.

Some philosophers think that the belief that human values are emotive and subjective is a by-product of modern individualism.[2] But since this idea can be found among the ancient Greek Sceptics, it is an unlikely explanation. More plausibly, a subjective view of ethics is a result of the hollowing out of religion. Expressed in universal laws or commands, 'morality' is a relic of monotheism. If there is no author of these edicts, what authority can they have? In religion, the author was God. Later, with the rise of the Enlightenment, it came to be 'humanity'. But humankind cannot be the author of anything, since there is no such thing as a universal human agent. All that exists is the multitudinous human animal, with its many different moralities.

For anyone reared to think that morality is single and universal, this is perplexing. So they continue to think and talk as if morality is clear to everyone, when in fact it is thoroughly opaque to those who practise it.[3]

SPINOZA ON LIVING ACCORDING TO YOUR NATURE

Fortunately there are other ways of thinking about the good life. In ancient Greece and China there were ethical traditions that made no reference to what is nowadays called morality. For the Greeks, the good life was living according to *dike* – your nature and its place in the scheme of things. For the Chinese, it meant living according to *tao* – the way of the universe, as manifest in your own nature. There are many differences between these ancient kinds of ethics. But it is what they have in common that is most useful to us today.

These ways of thinking do not feature 'morality', because neither assumes that one kind of life is dictated for all human beings by God. Nor do they assume that the core of a good life is concern for others. Instead, the good life means living for yourself with the nature you have been given. To be sure, the good life requires virtues – traits and skills that make it possible to survive and flourish – but these virtues concern not only what we have been taught to think of as morality. They also include aesthetics, hygiene and the whole art of life, and they are not confined to human beings. In this understanding, ethics – from *ethikos* in the Greek language, meaning 'character' or 'arising from habit' – is found among non-human animals.

That non-human animals possess virtues was recognized by Aristotle in the case of dolphins. At various points in his *History of Animals*, he notes how they suckle their young, communicate with one another and cooperate when chasing fish to eat.[4] He based these comments on direct observation while travelling with fishermen in the Aegean. Aristotle believed everything in the universe has a *telos* or purpose, which is to realize its nature as the kind of thing it is. A good life was one in which this was achieved. When dolphins came together to hunt for fish they were displaying a trait necessary to this end – in other words, a virtue. Dolphins were living the good life in a specific dolphin mode.[5]

Ancient Chinese thought contains a similar way of thinking. The Taoism of Lao Tzu and Chuang Tzu turned on the ideas of *tao* and *te*: the way or nature of things and the ability to live according to it. Though *te* has often been translated as 'virtue', it did not refer to any exclusively 'moral' capacity but the inner power that is needed in order to act according to the way of things. Following it meant acting as you must, and this was not only true for humans. All living creatures flourished only insofar as they obeyed their own natures.[6]

Aristotle's account of ethics is anthropocentric and hierarchical. Though he allows that virtues exist in other animals, he insists that the good life is realized most fully in a few human beings. The human mind most resembles God's – a divine intellect or *nous*, the final cause or 'unmoved mover' of the universe – and everything that exists is striving to be like God. It follows, for Aristotle, that the human animal is the *telos* – the end or goal – of the universe.

This idea fitted well with Christianity and persisted in popular theories of evolution. Darwin's theory, however, is altogether different: natural selection has no purpose and the human species has come about by accident. Humans are not 'better' than countless extinct species. But Darwin found it hard to stick to this view.[7] Today, many of his followers hold to the idea that humans are worth more than other animals, though this makes no sense unless you believe in a cosmic hierarchy of values.

In Taoist thought, on the other hand, human beings are in no way special. Like every other creature, they are straw dogs – ceremonial objects that were carefully prepared for rituals, then casually burned. As Lao Tzu puts it, 'Heaven and earth are ruthless, and treat the myriad creatures as straw dogs.'[8] The universe has no favourites, and the human animal is not its goal. A purposeless process of endless change, the universe has no goal.

In the central western tradition, humans rank higher than other animals because humans are more capable of conscious thought. For Aristotle the best life is intellectual contemplation of the cosmos, while for Christians it is the love of God. In both cases conscious awareness is an integral part of living well. For Taoists, in contrast, the self-regarding consciousness of human beings is the chief obstacle to a good life.

According to Aristotle, the best sort of human being was one like himself – male, slave-owning and Greek – who was devoted to intellectual inquiry. Apart from justifying the local prejudices of his time – a practice nearly universal among philosophers – this view has a more radical defect. It assumes that the best life for humans is the same, at least in principle, for everybody. True, most cannot achieve it, but this only shows their inferiority to those that can. The possibility that human beings might flourish in many different ways, which cannot be ranked in any scale of value, did not occur to Aristotle. Nor did the idea that other animals might live good lives in ways of which humans are incapable.

Taoism again offers a refreshing contrast. Human lives are not ranked in value and the best life for other animals does not mean becoming more like human beings. Each individual animal, every single creature, has its own form of the good life.

In western thought, the nearest approximation to this vision is Benedict Spinoza's idea of *conatus* – the tendency of living things to preserve and enhance their activity in the world. The neuroscientist Antonio Damasio, who finds in Spinoza (1632–77) anticipations of recent scientific discoveries regarding the unity of mind and body, cites a proposition in Spinoza's *Ethics* (1677) in order to clarify this idea:

> The quote comes from Proposition 18 in part IV of *The Ethics* and it reads: '. . . the very first foundation of virtue is the endeavor (*conatum*) to preserve the individual self, and happiness consists in the human capacity to preserve its self.'
> . . . A comment on the terms used by Spinoza is in order . . . First . . . the word *conatum* can be rendered as endeavor or tendency or effort, and Spinoza may have meant any of these, or perhaps a blend of the three meanings. Second, the word *virtutis* can refer not just to its traditional moral meaning, but also to power, and ability to act . . . here is the beauty behind the cherished quote, seen from today's perspective: It contains the foundation for a system of ethical behaviors and that foundation is neurobiological.[9]

The ambiguities to which Damasio refers are not incidental.

They testify to Spinoza's struggle to express a subversive philosophy in traditional terms.

There may be several reasons for Spinoza's equivocation. Like Montaigne he belonged to a Jewish family that fled the Iberian Peninsula to escape persecution and forcible conversion by the Inquisition. Bolder than Montaigne, he was expelled in 1656 from the central synagogue in Amsterdam for expressing some of the ideas which would be published posthumously in the *Ethics* to his co-religionists, who regarded them as heretical. Following his excommunication he was offered an academic position, which – fearing his freedom to think and write might be compromised – he refused. Instead he made a modest living as a lens-grinder, an occupation that may have shortened his life.

Spinoza's critics were right in regarding his views as heretical. For him, God is not the power that created the universe. God is an infinite substance, *Deus sive Natura*, God or Nature, self-subsisting and everlasting. Human values cannot be derived from a God that created the universe, since God is the universe itself. Spinoza's ambiguous language may have been an attempt to make this philosophy more palatable to those who had excommunicated him. But it may also be true that he underestimated the extent to which his philosophy undermined traditional beliefs. At times he seems to back away from what is most original in his thinking.

The British philosopher Stuart Hampshire, who reflected on Spinoza's philosophy for many years, explained the idea of *conatus*:

> Like every other identifiable particular thing in the natural order, a man tries in his characteristic activity to preserve himself and his own distinct nature as an individual, and to increase his own power and activity in relation to

his environment. This trying (*conatus*), or inner force of self-preservation, is that which makes any individual an individual . . .

. . . A man's natural tendency or *conatus* is not to make himself a good or perfect specimen of his kind, to realize in his activity some general ideal of humanity, but rather to preserve himself, this individual, as an active being, who is, as far as possible, independent in his activity. He has achieved virtue, and succeeded in that which he necessarily desires, when, and only when, he is comparatively free and self-determining in his activity.[10]

In Spinoza's view, 'good' is what furthers this endeavour, 'evil' what obstructs it. Values are not objective properties of things, but neither are they purely subjective. The virtue of an individual is what prolongs and extends their activity in the world. But the mass of human beings do not understand themselves or their place in the world. As a result, they are often mistaken in how they live.

In viewing good and evil in this way, Hampshire writes, 'Spinoza is representing the study of ethics, in the then dominant Christian and Jewish tradition, as one immense error, as the pursuit of a harmful illusion'.[11] The illusion comes partly from belief in free will. Traditional theories of morality suppose that various courses of action are open to us for choice and decision. But in Spinoza's account (as in some contemporary theories in neuroscience) what we think are our choices are the result of complex causes operating in our organism.[12] Our thoughts and decisions are not separate from our bodies, which function independently of what we take to be our conscious mind and will. The experience of deliberating and deciding to pursue an option is a by-product of our conflicting desires. Free will is the sensation of not knowing what you are going to do. In reality we are bound to maintain and extend our own power, though because of the fantasies that cloud the human mind this may not be achievable.

Spinoza believed that everything in the universe is as it must be. Nothing is accidental or contingent. That is why he rejected the idea of free will. But you need not accept Spinoza's metaphysical vision to grasp the power of his challenge to the traditional idea of morality. Nor do you have to accept that everything that exists is trying to keep on existing. All Spinoza's ethics requires is the idea that living things assert themselves as the particular organisms they are.

This is very different from the classical view, presented by Aristotle, in which everything strives to be a perfect specimen of its type, and from the monotheistic view, which says that humans achieve the good life by approaching the perfection of a divine being. Once you have given up these traditional beliefs, you will not be tempted by the idea that humans are unique in being able to choose their own good. You will think human beings are like other creatures in pursuing the good their natures demand.

Humans are driven by self-preservation. But because their minds are confused, they are often self-destructive. This could be remedied by second-order thinking:

> The central drive to self-preservation, which in part determines my other appetites, strictly matches a universal and unchanging feature of individual physical things. My rational reflections have as their objects my first-order desires and the activity of thought is embodied in a corresponding activity in the brain. In reflection, forming ideas of ideas, I evaluate the desire, or other thought, positively or negatively, either affirm it or deny it, or suspend judgement on it. The reflection

is an activity of the mind, its self-assertion, against the inputs from external things.[13]

In reality, no human being can achieve the freedom of mind Spinoza thought possible for a few. His idea that reflexive thought can clear the mind of fantasy is itself a fantasy. Spinoza wrote that his 'necessary truths' should be held onto 'as if they were rafts in a rough sea'.[14] But his axioms were figments, and his metaphysical life raft a leaky construction.

Even as he departed from a traditional conception of morality, Spinoza continued the rationalist tradition in which the most conscious life is the best. A divided mind could become one by identifying itself with the reason that is embodied in the cosmos. But if cosmic reason is a creation of the human imagination, reflexive thought – thinking about your thought processes – will only exacerbate inner division.

The flaw in rationalism is the belief that human beings can live by applying a theory. But theory – a term that comes from the Greek *theorein*, meaning 'to look at' – cannot replace practical knowledge of how to live. Plato misled western philosophy when he represented knowing the good in terms of visual experience. We can look at something without touching it; but the good life is not like that. We know it only by living it. If we think about it too much and turn it into a theory, it may dissolve and disappear. Contrary to Socrates, an examined life may not be worth living.

Spinoza renewed Plato's faith that the more conscious a life becomes, the closer it moves towards perfection. But if the value of a life is the value it has for the creature that lives it, any such hierarchy of value is without meaning. To live well does not mean being ever more conscious. The best life for any living thing means being itself.

This diverges from the Romantic view that each of us should fashion a unique individuality for ourselves. For the Romantics, human beings create their lives in the way artists do their works and the value of any work of art has to do with how original it is. Here the Romantics drew on a biblical idea of creation from nothing that is not found in ancient Greek thought. Romanticism is one of Christianity's many modern surrogates.

A Spinozist-Taoist ethic is quite different. Humans are like other animals. A good life is not shaped by their feelings. Their feelings are shaped by how well they have realized their nature.

For many people today, no way of living could be more oppressive. Contemporary culture rejects the idea of nature for the same reason it rejects the idea of God. Both set limits on the human will. Modern humanism follows the Romantics, who idealized the natural world but still regarded it as inferior to the best that humans could create. For these unwitting post-Christians, being free means rebelling against nature, including their own. For Spinoza and the Taoists, on the other hand, any such rebellion is self-defeating. Humans are like other creatures in trying to maintain and extend their power in the world. All of them are ruled by their *conatus*, the self-assertion of every living thing.

In Spinoza and Taoism, power means being able to be what you are. The languorous sloth asserts its power as it slumbers through its days as much as the tiger at the kill. Exercising power, in this sense, does not imply dominating others. But if ethics consists in the affirmation of your individual nature, you may find yourself outside of morality as understood by monotheists and humanists.

As an example, Spinoza considered pity a vice. 'Pity in a man who lives under the guidance of reason,' he writes in the

Ethics, 'is in itself bad and useless.' Pity is a kind of pain, he continues, and pain is evil. We may act to relieve the condition of someone we pity, but only if doing so is required by reason. Spinoza concludes:

> Hence it follows that a man who lives according to the dictate of reason endeavours as far as possible not to be touched with pity.
>
> He who rightly knows that all things follow from the necessity of the divine nature, and come about according to the eternal laws and rules of nature, will find nothing at all that is worthy of hatred, laughter, or contempt, nor will he feel compassion . . . he who is easily touched by the emotion of pity, and is moved to tears at the misery of another, often does something of which he afterwards repents; both inasmuch as we can do nothing according to emotion which we can certainly know to be good, and inasmuch as we are easily deceived by false tears.[15]

Spinoza's ethics diverges from traditional morality in not being composed of rules or laws handed down by some human or divine authority. It also views virtues and vices differently. Pity is a vice because it is a cause of sorrow and depletes vitality. Taoist ethics also departs from morality. Along with the way of the sage, it recognizes those of the tyrant and the assassin, the warrior and the criminal, together with that of the masses of human beings whose lives are spent resisting circumstance. Some are set on self-assertion, others bent on self-destruction. Some give life, others take it away. The ways of humans are ruthless, like the way itself.

This is far from the will to power that became a popular gospel in Europe in the late nineteenth and early twentieth centuries. In some of his later works, Friedrich Nietzsche

(1844–1900) flirted with the idea that everything in the world is a struggle for power. Nietzsche admired Spinoza, and claimed to have learned much from him, but Nietzsche's will to power is not the potency Spinoza finds in every particular thing. It is an inverted version of his early mentor Schopenhauer's universal will to live. The difference is that, whereas Schopenhauer laments the suffering the will to live brings, Nietzsche glories in the strife that it entails.

Before Nietzsche, the seventeenth-century English philosopher Thomas Hobbes asserted that humans were driven by an insatiable desire for power: 'I put for a general inclination of all mankind, a perpetual and restless desire of power after power, that ceaseth only in death.'[16] Hobbes thought this desire came from fear of other human beings, and more specifically of suffering a violent death at their hands. For humans, such a death was the *summum malum*, the supreme evil.

When he depicted human beings in a state of nature – a mythical construct denoting the absence of order in society – Hobbes came closer to human realities than his critics like to admit. War is as natural as peace, and history contains many periods in which violence is normal. Hobbes believed human beings could avoid this state of affairs by setting up a sovereign that would maintain order in society. But fear of a violent death is not the strongest human impulse. We do not live only to put off dying. Affirming your nature may mean courting death. Human beings willingly accept death as the price of protecting someone or something they love. Merely surviving is a wretched way to live, and it is not against nature to be ready to die. As we shall see in Chapter 5, human beings are also ready to die – and kill – for the sake of an idea with which they have identified themselves.

Spinoza's view of suicide is intriguing. Since everything tries to persist as the particular thing it is, no one could truly want to stop existing. No one wants to end their life: a suicide is someone who is killed by the world. As Spinoza puts it in the *Ethics*: 'Nothing can be destroyed save by an external cause.'[17] From another point of view, people commit suicide when their *conatus* turns against itself.

Spinoza believed that human beings – if they were fully rational – could avoid thinking of death at all. In a famous passage of the *Ethics*, he writes: 'A free man thinks of nothing less than of death, and his wisdom is a meditation not of death but of life.' He believed he could demonstrate the truth of this proposition:

> *Proof.* – A free man, that is, one who lives according to the dictate of reason alone, is not led by fear of death . . . but directly desires what is good . . . that is . . . to act, to live, and preserve his being on the basis of seeking what is useful to him. And therefore he thinks of nothing less than of death, but his wisdom is a meditation of life. *Q.e.d.*[18]

Spinoza's QED is fanciful. We can repress the thought of death as we do many of our thoughts, but that only leaves it lurking in a darkened part of the mind. The human being who thinks nothing of death does not exist.

SELFLESS EGOISM

We inherit the belief that morality in its highest form means altruism – that is, selflessness or living for others. Empathy, in this tradition, is the heart of the good life. Cats, on the other hand – except where their kittens are concerned – show few signs of sharing the feelings of others. They may sense when their human companions are distressed, and stay with them through a time of trouble. They may give succour to the sick and dying. But cats are not sacrificing themselves in any of these roles. Just by being there, they are giving human beings relief from sorrow.

As predators, a highly developed sense of empathy would be dysfunctional for cats. That is why they lack this capacity. It is also why the popular belief that cats are cruel is mistaken. Cruelty is empathy in a negative form. Unless you feel for others, you cannot take pleasure in their pain. Humans displayed this negative empathy when they tortured cats in medieval times. In contrast, when cats toy with a captured mouse they are not revelling in its torment. Teasing their prey expresses their nature as hunters. Rather than torturing creatures in their power – a singularly human predilection – they are playing with them.

The link between altruism and the good life may seem self-evident, but it is a novelty in ethics. Caring for others does not feature much in ancient Greek values. There is nothing in Aristotle about self-sacrifice; the 'great-souled man', when not contemplating the cosmos, spent his time admiring himself. Nor is altruism very prominent in early Buddhism, where the goal is to shed the illusion of the self in order to achieve a state of complete quietude or *nirvana*. The Buddha seems to have believed that only humans could achieve this liberation. Earlier Indian traditions forming part of the body of practices now called Hinduism believed each creature achieved freedom in acting according to its nature. Here, Hindu traditions are closer to Taoism than to Buddhism. For the Buddha, liberation meant renouncing selfhood, but the goal was still to liberate oneself. Only later in the history of Buddhism did the idea arise that in a supreme act of selflessness an awakened individual (*bodhisattva*) might renounce *nirvana* to be reborn to liberate all sentient beings.

In the case of Christianity, the good life has not always meant helping others. The nineteenth-century Russian religious philosopher Konstantin Leontiev, who in later life entered an Orthodox monastery and died as a monk, saw Christianity as a type of 'transcendental egotism' – a way of life focused on individual salvation.[19] Christianity is commonly described as a religion of love, but the love of which Christian mystics speak is love of God. Human beings receive love as God's children, but if they err they risk damnation. The Christian religion is no more defined by universal love than is Buddhism.

Today there are philosophers who think the best life is the one that does the most good. Developing a Utilitarian philosophy promoted by nineteenth-century thinkers such as Jeremy Bentham, they believe the best life maximizes general welfare – often defined as satisfying the desires of everyone affected by one's actions. It does not occur to these exponents of 'effective altruism' to question whether living the good life and doing the most good are one and the same.[20] After all, it is only an historical accident that they are equated today. Few would think this way if Christianity had not triumphed and the West was still ruled by some version of Graeco-Roman ethics.

Altruism is a modern idea. The word was coined by the French sociologist Auguste Comte (1798–1857) to define the core of the Religion of Humanity he invented and propagated. In this supposedly scientific religion, a good life was one that served 'humanity', not any divine being. To be sure, the altruism he advised his disciples to practise was not directed to any actually existing human being. The beneficiary – the enlightened species he believed was emerging – was as much a creation of the human imagination as the Deity it replaced, and if anything more incredible.

Though nowadays nearly forgotten, Comte's secular religion had an enormous influence in equating morality with altruism. In recent decades, hundreds of books have appeared that argue that morality can be explained in evolutionary terms. All of them take for granted that moral behaviour is essentially altruistic, an assumption that is historically and culturally parochial: the Christian-humanist conception of the good life as living for others is only one of many in which human beings have found fulfilment.

Yet so deeply has this conception entered into popular and scientific thinking that ethics has been described as being 'aimed at remedying disruptive altruism failures'.[21] In biology, altruism is a concept that refers to cooperative behaviour, mostly within groups. Having shown that altruism has evolutionary functions, some philosophers believe they have explained ethical life among humans. But if they have explained anything it is only a watery version of Christian morality, reformulated in the faux-Darwinian terms of the modern secular intellectual.

Neither Spinoza nor Taoism thinks about the good life as living for others. At the same time they link self-realization with a kind of egolessness. As pointed out by Paul Wienpahl, an American philosopher and scholar of Spinoza who was also a long-standing practitioner of Zen meditation, the seventeenth-century French philosopher and sceptic Pierre Bayle identified this affinity between Spinozism and Zen Buddhism. Wienpahl writes:

> The first notice, that I know, of a resemblance between Benedict Spinoza and Ch'an or Zen Buddhism occurs in the entry on Spinoza in Pierre Bayle's dictionary, where Bayle

related Spinozism to 'the theology of a sect of the Chinese'. It is clear when you read his description that he was reporting on Jesuit accounts of Ch'an Buddhists, or, as the Jesuits called them, the followers of Foe Kaio (No Man). In this notice Bayle said there was nothing new in Benedict Spinoza, for 'the theology of a sect of the ancient Chinese' is also based on the incomprehensible notion of nothingness . . . For Bayle this means that everything substantial is removed from reality.[22]

Wienpahl goes on to observe that a central notion in Zen is the nothingness of the self. The Zen school emerged in China as a result of an interaction between Buddhism and Taoism, and the insight that human selfhood is illusory is common to both of them.

Another scholar who recognized an affinity between Spinozist and Taoist ethics was the Norwegian Jon Wetlesen. In *The Sage and the Way: Spinoza's Ethics of Freedom*, Wetlesen writes that Taoism 'does not aim at becoming what one is not, but in being what one is. This does not require any special doings on the part of the temporal ego, but rather an undoing of the ego.'[23] He finds the same distinction between the ego and the true nature of the individual in Spinoza.

An ethics in which you realize your individual nature differs from any idea of self-creation. The self with which humans identify is a construction of society and memory. Forming an image of themselves in infancy and childhood, they seek happiness by preserving and strengthening that self-image. But the image they have of themselves is not the reality of their bodies or their lives, and chasing after it can lead not to fulfilment but to self-frustration.

Other animals do not share their lives with any such phantom. Most lack any image of themselves. Self-preservation for them means not the continued existence of an imagined self but the ongoing vitality of the body. They are not shadowy selves examining their thoughts and impulses as if they belonged to some other being. When they act, it is without the human sense that it is a separate entity in themselves – a mind, a self – that is acting.

In their lack of a deceptive self-image, cats are exemplary. They are not among the select group that has passed the Mirror Self-Recognition (MSR) test developed in 1970 by the American psychologist Gordon Gallup Jr. The test requires that the animal be given some physical mark, often a dot of colour, on a part of the body that is only visible in a mirror. If the animal tries to touch the part of their body that displays the colour, they are deemed to have shown self-awareness. Humans, chimps, bonobos and gorillas passed the test, along with cetaceans such as dolphins and orca whales, and some birds such as the magpie. Other corvids, pigs and macaques have demonstrated partial self-awareness in the test.

Cats respond to their own reflection with indifference, or else react to it as if it were another cat. Some cats have been reported as being distressed when humans laugh at them, and some breeds – Siamese, for example – are reputed to be vain. But rather than being upset at how they are perceived, these cats may be interpreting the human response to them as hostile or dangerous. Again, cats may be coquettish or menacing in the company of other cats, but they are not burnishing an image they have made of themselves. They are projecting an image of themselves to other cats for the sake of courting a mate or protecting their territory.

Studies show that cats may recognize their names, but when they are called may not care to respond.[24] Their history of interacting with humans has not left them so dependent that they need to answer to the name humans know them by. Unlike dogs, they have not acquired any of the human sense of self. Certainly they distinguish themselves from a world that is outside of them. But it is not an ego or a self in them that interacts with the world; it is the cat itself.

Feline ethics is a kind of selfless egoism. Cats are egoists in that they care only for themselves and others they love. They are selfless in that they have no image of themselves they seek to preserve and augment. Cats live not by being selfish but by selflessly being themselves.

Traditional moralists will resist the very idea of feline ethics. How can a creature be moral if it cannot grasp principles of right and wrong? Surely only behaviour that is enacted to obey such principles can be moral. Action must have a reason that can be known to the actor, otherwise there can be no morality.

It is a familiar refrain. But if this is what morality requires, humans cannot be moral either. True, they may come up with some principle or other and then try to stick to it. But they have very little idea why they act as they do. Why one principle, and not another? If two or more principles conflict, how can they decide between them? If they find a reason for acting as they do, how do they know it has moved them to act? Human beings no more choose to act 'morally' than they choose to sneeze or yawn. Philosophies in which a good life consists of self-chosen behaviour are sleights of hand whose purpose is to fool the conjuror.

The mistake is the belief that that a good life is one that pursues an idea of the good. An idea here means a kind of vision, as in Plato. Having glimpsed the good, we spend our lives struggling to approach it. Cats do nothing of the kind, of course. Though they can see in the dark, scent and touch are more important in their lives. A good life is one they have felt and smelt, not a dim sighting of something far away.

A good life need not embody any idea. Someone who responds to the suffering of others by helping them displays compassion whether or not they have any idea of what they are doing. They may be more truly virtuous if they have no notion that they are being compassionate. The same goes for courage.

As in human beings, the good life in cats depends on their virtues. Aristotle pointed out that someone lacking in courage cannot flourish no matter what other virtues they may possess: whatever they do will come to nothing. Similarly, a cat that is always fearful cannot live well. Whether in the wild or in human settlements, a cat's life is dangerous. Courage is as much a feline virtue as it is a human one. Without it, neither cats nor humans can thrive.

A good life for any living thing depends on what it needs to fulfil its nature. The good life is relative to this nature, not to opinion or convention. As Pascal observed, human beings are unusual in having a second nature formed by custom, along with the nature they have when they are born. It is natural for them to mistake their second nature for the first, and many who lived by the customs of their societies have lived badly as a result. Mistaking their own nature is not a habit of cats.

To be sure, we cannot know what it is like to be a cat. Nor can we know what it is like to be another human being. Yet we rightly think that someone who believes other human beings are unfeeling machines has a mental illness, whereas philosophers such as Descartes who thought the same of other animals have been feted as sages. In fact the inner world of cats may be more lucid and vivid than our own. Their senses are sharper and their waking attention unclouded by dreams. The absence of a self-image may make their experiences more intense.

Judging by the single-minded way in which cats conduct themselves, the feline condition of selflessness has something in common with the Zen state of 'no-mind'. One who achieves 'no-mind' is not mindless. 'No-mind' means attention without distractions[25] – in other words, being fully absorbed in what you are doing. In humans this is rarely spontaneous. The supreme archer is one who shoots the arrow without thinking, but this comes only after a lifetime of practice.[26] Cats have no-mind as their inborn condition.

Philosophers who deny conscious awareness to other animals ascribe to themselves a state of mind they know only fitfully, if at all. The inner life of humans is episodic, fuzzy, disjointed and at times chaotic. There is no self that is more or less self-aware, only a jumble of experiences that are more or less coherent. We pass through our lives fragmented and disconnected, appearing and reappearing like ghosts, while cats that have no self are always themselves.

4

Human vs Feline Love

The passionate attachments of love are at the centre of many human lives. Mostly it is love of another human being, but it may also be love of a non-human animal. On occasion these loves may come into conflict. Literature and memoirs can illuminate the differences between the two.

SAHA'S TRIUMPH

A collision of human with feline love is the subject of *The Cat* (1933), a short novel by Sidonie-Gabrielle Colette (1873–1954). Born into a declining middle-class family, the French author was induced, at the age of twenty, to marry a well-known writer who then used her literary talents to publish a number of novels under his own name. In 1906 Colette left her husband and spent some years earning an uncertain living as a stage performer. In 1912 she married the editor of a national newspaper, but partly as a result of an affair she had with her sixteen-year-old stepson the two were divorced twelve years later. She married again in 1925, and the marriage lasted until she died. She also had loving relationships with women, some of them extending over many years; and had a passion for cats, which she described as being necessary to her solitude. She never stopped writing for long, and in 1948 was nominated for the Nobel Prize in Literature. By the time of her death she was one of the most highly respected writers in the world.

Colette's biographer Judith Thurman writes of *The Cat*: 'In this novel of which a cat is the romantic heroine, Colette's prose is particularly feline – both detached and voluptuous, minutely observant of those pleasures and irritants of the flesh which are lost on grosser human senses.'[1] The heroine is Saha, a golden-eyed little Chartreux (Russian Blue) who lives with Alain, a dreamy young man who enjoys more than anything else spending time with Saha in the beautiful garden of the crumbling family villa. Encouraged by his mother, Alain marries Camille, a sexually uninhibited young woman of nineteen, and the two 'fell into the diversion which shortens the hours and makes the body attain its pleasure easily'.[2] But Alain quickly tires of Camille. Her body seems less beautiful than he imagined, and he is worn out by her demands for sex. It is not long before he is bored with her. Whenever he can, he retreats to the garden with Saha.

Camille grows more jealous, and one morning when Alain is away she throws Saha out of the window of the high-rise apartment in which they are living. Saha's fall is broken by an awning, and she survives unhurt. The unsuccessful attempt to kill his cat enables Alain to free himself from a human connection that has become oppressive. Carrying Saha in a basket, he returns to his mother. Next morning Camille appears, asking for forgiveness. Alain will have none of it. Speaking slowly and quietly, he tells her: 'A little blameless creature, blue as the loveliest dreams. A little soul. Faithful, capable of quietly, delicately dying if what she has chosen fails her. You held *that* in your hands, over empty space . . . and you opened your hands. You're a monster. I don't wish to live with a monster.'[3]

Camille is horrified at being 'sacrificed' for the sake of an animal. After some angry exchanges the two part, leaving their future unsettled. Exhausted, Alain sinks into a chair. Suddenly, 'like a miracle', Saha appears on a wicker table beside him. 'For a while Saha, on guard, was following Camille's departure as intently as a human being. Alain was half-lying on his side, ignoring it. With one hand hollowed into a paw, he was playing deftly with the first green, prickly August chestnuts.'[4]

These last words in the novella encapsulate its central theme. Loving Saha more than he does any human, Alain himself becomes cat-like. Though she is portrayed obliquely, the Russian Blue is the most fully realized character in the story. Camille's jealousy is nakedly revealed; Saha's is only hinted at. The triumph of the cat is clear from the start.

For a cat-lover it is a delightful tale. The flaw in the story is Saha's own jealousy. Cats may be jealous of other cats, though what seems to us jealousy may be no more than a reaction to the disruption of their habits that comes when another cat enters their territory. Cats seldom show jealousy when another human being comes into the life of the human they live with. Dogs may demand the exclusive attention and devotion of their masters. In his memoir *My Dog Tulip* (1956), the British author, editor and broadcaster J. R. Ackerley recalls the intense possessiveness of his canine companion Queenie.[5] (The dog's name was changed in the book on the grounds that it could be seen as pointing to the author's sexuality, though Ackerley never concealed the fact that he was gay.) Ackerley's book is one of the great accounts of love between a human and a non-human animal, but it could not have been written if Queenie had been a cat.

Anyone who has lived with cats knows they can enjoy being with us. When they lie on their back and ask to be tickled, they are exposing the most vulnerable part of their body to a human for whom they have come to feel trust and affection. They take pleasure in our company and in playing with us. But this is not any exclusive attachment of the kind Ackerley describes with Queenie. Cats often have several homes, each chosen by them, to which they go for food and attention. If its closest human companion leaves for a time, a dog will be distressed. A cat may seem hardly to notice when the most familiar human in its life goes away. Cats may come to love human beings, but that does not mean they need them or feel any sense of obligation to them.

MING'S BIGGEST PREY

The American novelist and short-story writer Patricia Highsmith (1921–95) invented the amoral killer Tom Ripley, who features as the central figure in five of her books and a number of movies based on them; she also wrote stories in which mistreated animals take revenge on human beings. Highsmith's biographer Andrew Wilson writes of these tales: 'By positioning the animals as subjects, by giving voice to their thoughts, Highsmith disrupts the Western philosophical tradition which celebrates the rationalism of man.'[6] One story features a cockroach, who feels he is as entitled to call himself a resident of the hotel in which he lives as the humans who stay there.

Some have claimed Highsmith modelled Ripley on her cats and it has been reported that she called one of them Ripley after her psychopathic anti-hero. But only humans can be psychopaths.[7] Cats may at times seem impassive, but that is only because they express emotion through their ears and tails rather than their faces. They also express their feelings through purring. Usually purring is a sign that they are happy, but not always – sometimes it can signal distress. Either way, there is no deception in it.

Highsmith's sympathy with non-human creatures was deeply felt. When walking in Soho she came across a wounded pigeon lying in the gutter; her companion persuaded her it could not be rescued, but she was visibly distressed. She was horrified by the cruelty of battery chicken farming, and said that if she could discover who docked the tail of a local black cat she would not hesitate to shoot them – 'and to kill'. She was extremely fond of snails, breeding them in her garden in Suffolk and on occasion carrying over a hundred of them, together with an enormous head of lettuce, in her handbag. When she moved to France, she smuggled in some of her pet snails by hiding them under her breasts.[8] Her carer in old age recalled that she would return spiders that had strayed into her home back into the garden, making sure they were not damaged. 'For her human beings were strange – she thought she would never understand them – and perhaps that is why she liked cats and snails so much.'[9] A long-time friend wrote of her: 'As for animals in general, she saw them as individual personalities often better behaved, and endowed with more dignity and honesty than humans.'[10]

Struggling with her sexuality as a young adult, Highsmith had a course of psychoanalysis with a therapist who tried to 'cure' her of being gay. For a time she seems to have considered entering into a conventional marriage. She went on to have many female lovers and some enduring friendships with gay men, but appears not to have found the companionship she enjoyed with animals. She was passionate in her love of cats, writing that they 'provide something for writers that humans cannot: companionship that makes no demands or intrusions, that is as restful and ever-changing as a tranquil sea that barely moves'.[11]

In 'Ming's Biggest Prey', Highsmith has a beautiful Siamese take revenge on her mistress's lover. Ming preferred a quiet life:

> Ming liked best lying in the sun with his mistress on one of the long canvas chairs on their terrace at home. What Ming did not like were the people she sometimes invited to their house, people who spent the night, people by the score who stayed up very late eating and drinking, playing the gramophone or the piano . . . People who stepped on his toes, people who sometimes picked him up from behind before he could do anything about it, so that he had to squirm and fight to get free, people who stroked him roughly, people who closed a door somewhere, locking him in. *People!* Ming detested people. In all the world, he liked only Elaine. Elaine loved him and understood him.[12]

When Elaine's new lover Teddie tries to push Ming overboard during a sailing trip off the coast of Acapulco, Ming decides to hit back. Later in the same day, when they have returned to the villa, Teddie tries to get rid of Ming again, this time by throwing him over the terrace. Ming jumps on his shoulder, and the two of them fall to earth together. Teddie is killed, while Ming is only out of breath. Recovering from the struggle, he settles himself with his paws tucked under him in a shadowy corner of the terrace, which was still warm from the sun:

> There was a great deal of talk below, noises of feet, breaking of bushes, and then the smell of all of them mounted the steps, the smell of tobacco, sweat, and the familiar smell

of blood. The man's blood. Ming was pleased, as he was pleased when he killed a bird and created this smell of blood under his own teeth. This was big prey. Ming, unnoticed by any of the others, stood up to his full height as the group passed with the corpse, and inhaled the aroma of his victory with a lifted nose.[13]

The story ends with Ming and his mistress together in her bedroom. Elaine strokes Ming's head, lifts his paw and presses it gently so that the claws come out. '"Oh, Ming – Ming," she said. Ming recognized the tones of love.'[14]

Like Colette's, this is a pleasing tale for cat-lovers. The story is told from Ming's point of view, and he is an attractive figure throughout. He may have no special fondness for humans in general, but he decides that Teddie is an enemy only after Teddie tries to kill him, and, when Ming takes his revenge, what he does could just as well be called self-defence. Ming's relationship with Elaine is harder to decipher. There is no doubt she loves him. It is left open whether he reciprocates her affection, or regards her simply as a creature he is happy to be with. And if the latter, might that not be love too?

Feline love differs from human love for many reasons. Sexual contact between male and female cats lasts a few moments, and is not followed by a life together. Except in the case of lions, who may protect their cubs, male cats take no part in bringing up their offspring. As soon as the kittens have learned the necessary skills from their mother, they leave for life on their own. But love among cats has qualities many kinds of human love lack. Cats do not love in order to divert themselves from loneliness, boredom or despair. They love when the impulse takes them, and are in company they enjoy.

In her youth Highsmith was a devotee of Marcel Proust – the supreme analyst of human love. Until her mid-twenties she thought of herself as being as much an artist as a writer, and throughout her life continued to produce paintings, drawings and wood-carvings. After her death a number of her drawings were published, many of cats, together with one entitled *Marcel Proust Examining His Own Bathwater*.[15]

In Proust's work, human love is forensically dissected. The Proust scholar Germaine Brée writes:

> Society forms a sort of biological culture in which individuals can try out all the means by which they can make contact with one another . . . Love is born there . . . But, more than anything else, what flourishes in all its forms is the need for 'diversion' in the Pascalian sense of the word.
>
> 'Diversion' for society people is the art of using others for the sole purpose of satisfying one's needs and disguising one's boredom. Where affections are concerned, this exploitation can be admitted neither to oneself nor to others. That is why Proust's characters hide, dissemble, and betray one another. They lie to themselves and to each other under various pretexts, hiding their real motives . . . Possessing both money and leisure, left entirely to its own devices, the social set has only one deep-rooted desire: to be protected from the emptiness of existence and to draw from the sterile and disquieting substance of life a mask which is reassuring and flattering to itself . . . They want neither to understand nor to know, only to be adorned and amused.[16]

As Brée indicates, Proust's analysis of love has much in common with Pascal's account of diversion. Where Proust differs from Pascal is in thinking that diversion obeys impersonal laws. Love is the product of mechanisms of which the lover knows nothing, and the charade of infatuation and disillusion shows them in the grip of forces they cannot comprehend or control. Vanity and jealousy impel them into an imaginary world, where they can forget their ageing bodies and the path to death on which they are bound. Erotic love is the working of a machine, and the mechanical quality of this love is its saving power. Even the most intense jealousy and the bitterest disappointment give a temporary respite from emptiness. Love erects a barrier against knowledge, against understanding – whether of others or oneself – that allows human beings relief from being themselves.

In this Proustian analysis, human love is more mechanical than the coupling of beasts. In love, more than anywhere else, human beings are ruled by self-deception. When cats love, on the other hand, it is not in order to fool themselves. Cats may be egoists but they do not suffer from vanity – not in regard to humans, at any rate. What they want from humans is a place where they can return to their normal state of contentment. If a human being gives them such a place, they may come to love them.

LOVING LILY

The novelist Junichirō Tanizaki (1886–1965) was celebrated for portraying the transformation in Japanese life that took place with the country's modernization. Much of his work asks what may have been lost along the way. One thing that was lost, he believed, was a distinctive sense of beauty. In a long essay, *In Praise of Shadows* (1933), Tanizaki wrote:

> we find beauty not in the thing itself but in the patterns of shadows, the light and the darkness, that one thing against another creates. A phosphorescent jewel gives off its glow

and color in the dark and loses its beauty in the light of day. Were it not for shadows, there would be no beauty.[17]

It is not that Tanizaki prefers darkness to light. Darkness is part of light's beauty:

We do not dislike everything that shines, but we do prefer a pensive luster to a shallow brilliance, a murky light that, whether in a stone or an artefact, bespeaks a sheen of antiquity ... Yet for better or for worse we do love things that bear the marks of grime, soot, and weather, and we love the colors and the sheen that call to mind the past that made them.[18]

A feature of this aesthetic is its distaste for perfection. A strand in western aesthetics cannot help thinking of beautiful things as flawed embodiments of an immaterial idea. Plato's mystical vision has led western philosophers to think of beauty as an otherworldly radiance. In contrast, Tanizaki writes of 'the glow of grime'.[19] True beauty is found in the natural world and everyday life.

Tanizaki was interested in the varieties of love and what they reveal of human beings. One of his most delicate explorations of this theme appears in his novel *A Cat, a Man and Two Women*, first published in 1936 and later made into a film, in which the central character is an ageing but elegant tortoiseshell called Lily.

The story begins with a letter about Lily written by one of the women to the other. Shinako begs her former husband Shozo's new wife Fukuko to give her the cat:

There's just one thing I want from you. And of course by that I *don't* mean I want you to return *him* to me. No, it's something much, much more trivial than that. It's Lily I want ... Considering all I've sacrificed, is it too much to ask for one little cat in return? To you it's just a worthless little animal, but what a consolation it would be to me! . . . I don't want to seem like a crybaby, but without Lily I'm so lonely I can hardly stand it . . . Why, there's nobody in this whole world who'll have anything to do with me now, except for that cat . . .

It's not you who won't give Lily up, but *him*. Yes, I'm sure of it. He loves her. 'I might be able to do without you,' he used to say, 'but do without Lily? Never!' And he always paid much more attention to her than he did to me at the dinner table, and in bed . . . do be careful, Fukuko dear. Don't think, 'Oh, it's just a cat,' or you may find yourself losing out to it in the end.[20]

At first reading, the story tells of three human beings using a cat as a weapon in their conflicts with each other. The cat seems to be a pawn in the family conflicts. But Lily means more to them than they realize. When the cat has gone to Shinako, Shozo has an overpowering sense of loss. He decides to go to Shinako's home secretly to see Lily. Crouching in a clump of bushes outside his ex-wife's house, he notices a plant from which an occasional gleam comes. 'Shozo's heart leapt within him at each gleam, hoping that it might be the glow of Lily's eyes: "Is that her? Wouldn't it be *wonderful*!"' His heart beat faster, and there was a chill in the pit of his stomach:

> Odd as it may sound, Shozo had never experienced this sort of agitation and impatience before, even in his relations with other human beings. Fooling around with café waitresses was all he'd been capable of. The closest he had come to a love affair was when he was seeing Fukuko on the sly, hiding it from Shinako . . . Even so . . . his affair with Fukuko had

always lacked a certain seriousness: never had his desire to see her or meet her been anything like as intense as his feelings for Lily were now.[21]

Fearing a row with his wife if he is late home, Shozo leaves. But he has not given up trying to see Lily. Next day he returns to his ex-wife's house. Shinako has gone out, leaving her sister Hatsuko in charge. Hatsuko takes Shozo up a steep flight of stairs to the room where Lily was resting. With the curtains closed, the room was shaded, but he could make out Lily sitting on a pile of cushions, her front paws folded under her, her eyes half-closed. The sheen of her coat showed she was well looked after, and some rice and an eggshell nearby that she had just finished her lunch.

Shozo was grateful Lily was safe and well. He smelt the cat's litter, and a loving sadness overwhelmed him. 'Lily!' he cried. The cat made no response. Then, 'seeming to notice his presence at last, [the cat] opened two dull, listless eyes and cast an extremely unfriendly glance in Shozo's direction. Apart from that, there was no expression of emotion. Folding her forepaws still more deeply under her, and twitching the skin on her back and at the base of her ears as if she were cold, she closed her eyes again with a look that expressed the need for sleep, and sleep alone.'[22]

Shozo tried stroking her, but Lily just sat with her eyes shut, purring. Shinako must care for the cat deeply, Shozo reflected. She was poor now, but despite her poverty she was making sure Lily was well fed. Lily's cushions were thicker than Shinako's own. Then he heard the sound of footsteps, and realized Shinako had returned. Shozo scuttled down the stairs and into the street, just missing her. The story ends: 'As if pursued by something dreadful, he ran at full speed in the opposite direction.'

Lily may have been used as a weapon by the human beings in her life, but she was the only one of the four of them to have been loved. Shozo and his ex-wife cared for Lily more than they cared for each other, or perhaps any human being. Their stratagems with regard to one another were outdone by the love each of them felt for the cat. Maybe this love was like much love that exists between humans: a refuge from unhappiness. Or maybe it was love for the cat itself; a mix of tenderness and admiration. What she felt about the humans in her life cannot be known. By the end of the story she has aged, and wants most of all to sleep. She may sense she is approaching death. Yet Lily is still the light in the room, and the human beings dim figures in the glowing prism of her mind.

GATTINO VANISHES

A different love between a human being and a cat is described in Mary Gaitskill's exquisite essay 'Lost Cat'.[23] Gaitskill's essay differs from the other stories recounted in this chapter: like Jack Laurence's story of Mèo in Chapter 1, it is a memoir of the life and death of an actual cat.

Born in 1954, Gaitskill became a literary celebrity with a collection of short stories, *Bad Behavior*, published in 1988. She had struggled both personally and financially for many years. As a teenager she was expelled from boarding school and committed by her parents to a mental institution from which she ran away. As a young adult she worked as a flower-seller, stripper, bookstore assistant, night-time proof-reader, freelance fact-checker and other casual occupations.

At one point she lived in a cheap sublet above a well-known

New York S&M Club, and many of her stories deal with the human need for pain and humiliation. One of them became the successful movie *Secretary* (2002), though Gaitskill felt it 'too cute and ham-fisted'.[24] In a later novella, *This is Pleasure* (2019),[25] she recounts how a dandyish book editor who enjoys evoking a need for pain and punishment in women is professionally destroyed when he is accused of sexual assaults on women he employed.

A recurring theme of Gaitskill's work is the contradictoriness of human love. Humans look to love for relief from boredom, the comfort of being an object of affection or obsession, an opportunity to wield power and inflict pain on themselves and others, and the excitement that can come from self-destruction. Love between humans and animals lacks these blemishes, and losing it may be more shattering than the end of a purely human love.

In 'Lost Cat', Gaitskill tells how she lost her cat when he was only seven months old. She had found the cat while visiting a literary aristocrat in Tuscany who had turned her estate into a writer's retreat. One of three skinny kittens in the yard of a nearby farmhouse, more sickly-looking than the other two, he tottered up to Mary. His eyelids were almost glued shut with mucus. A soft grey tabby with black stripes, 'He had a long jaw and a big nose shaped like an eraser you'd stick on the end of a pencil. His big-nosed head was goblinish on his emaciated potbellied body, his long legs almost grotesque. His asshole seemed disproportionately big on his starved rear. Dazedly he let me stroke his bony back; tentatively, he lifted his pitiful tail.'[26] Later she would remember him, 'back arched, face afraid but excited, brimming and ready before he jumped and ran, tail defiant, tensile, and crooked . . . Even if he was weak with hunger. He had guts, this cat.'[27]

He had only one half-good eye, and Mary called him Chance. 'I liked Chance, as I like all kittens; he liked me as a food dispenser. He looked at me neutrally, as if I were one more creature in the world.'[28] After some time, Chance began to raise his head when Mary came into the room and look at her intently. 'I can't say for certain what the look meant; I don't know how animals think or feel. But it seemed that he was looking at me with love. He followed me around my apartment. He sat in my lap when I worked at my desk. He came into my bed and slept with me; he lulled himself to sleep by gnawing softly on my fingers. When I petted him, his body would rise into my hand. If my face was close to him, he would reach out with his paw and stroke my cheek.'[29]

Mary's husband didn't like the name Chance, and Mary wasn't sure about it either, so they called the kitten McFate instead. McFate's strength increased, and he acquired 'a certain one-eyed rakishness, an engaged forward quality to his ears and the attitude of his neck that was gallant in his fragile body. He put on weight, and his long legs and tail became soigné, not grotesque. He had strong necklace markings on his throat; when he rolled on his back for me to pet him, his belly was beige and spotted like an ocelot's. In a confident mood, he was like a little gangster in a zoot suit.' Yet he was still delicate, and Mary decided that McFate was 'too big and heartless a name for such a small fleet-hearted creature', so she called him Gattino.[30] Mary told her husband she wanted to take Gattino back to the US with them. He was bemused, as she knew many people would be. They 'would consider my feelings neurotic, a projection onto an animal of my own need'.[31]

When Mary decided to take Gattino back home with her, she asked herself if it was wrong to love an animal more than suffering human beings. She had loved human beings, including children she had hosted through a programme for poor urban black families, and her father, who died a painful death from cancer after refusing treatment. But these loves had been tangled and frustrating:

> Human love is grossly flawed, and even when it isn't, people routinely misunderstand it, reject it, use it, or manipulate it. It is hard to protect a person you love from pain, because people often choose pain; *I* am a person who often chooses pain. An animal will never choose pain; an animal can receive love far more easily than even a very young human. And so I thought it should be possible to shelter a kitten with love.[32]

Sometimes when walking near the aristocrat's estate Mary would think of her father. During her walks, she carried a large marble that had belonged to him. Without really believing it possible, she wondered if part of her father's soul might have been reborn in Gattino. One night, when the kitten was lying purring in her lap, Mary saw a small sky-blue marble rolling on the floor from under a dresser. 'It was beautiful, bright, and something not visible to me had set it in motion. It seemed a magical and forgiving omen, like the presence of this loving little cat.' She put it on the windowsill next to her father's marble.[33]

As part of getting a pet passport, Mary took Gattino to the vet, who placed him in a cage near an enormous dog, which growled and barked at the kitten. At first Gattino hid behind a little bed, then defiantly faced the dog. '[T]hat is when I first saw that terrified but ready expression, that willingness to meet whatever was coming, regardless of its size or its ferocity.'[34] When she made the long transatlantic journey home, Gattino accompanied her, 'peering intrepidly from his carrier. And Gattino *was* intrepid. He didn't cry in the car, or on the plane, even though he'd had nearly nothing to eat since the night before. He settled in patiently, his slender forepaws stretched out regally before him, watching me with a calm, confidently upraised head . . . If I'd let him, he would've wandered up and down the aisles with his tail up.'[35]

Once in the US, Gattino was introduced to the other cats in the house. Approaching them with respectful tact, he settled in well. The family moved to a new house, where the landlord had left junk everywhere and the stove was broken and full of mouse nests. Things started to go wrong. Mary lost her passport; her husband mislaid a necklace she had given him. She also lost the blue marble she had found in Italy. But Gattino loved their new home. He played in the yard with the other cats, and showed no interest in going out into the street – and if he did, Mary thought he would soon find his way back as there was an open field across the road.

This is when Gattino disappeared. Returning home after a few hours, Mary looked everywhere for him in the dark. It was at this point that some words first entered her head: 'I'm scared.' Gattino was attuned to her, she felt. She wanted to respond, 'Don't worry. Stay where you are. I will find you.' But instead she thought, 'I'm scared, too. I don't know where you are.' She worried that sensing her fear would make him feel more lost, but she could not help herself. She put up posters, sent out emails and alerted campus security at the nearby university. Three nights later, another thought came to her: 'I'm lonely.' On the fifth night, she received a call from a guard saying he had seen a small, thin, one-eyed cat foraging in a garbage can. The call came at two in the morning, when the phone was turned down. Mary and the family didn't hear it.

Then Mary decided to see a psychic who had been recommended by a friend. The psychic told her Gattino was in trouble and dying. She described the place where Gattino might be, and Mary went and searched there for several days and nights. At the end of one of the nights, when she was about to fall asleep, words again formed in her mind: *I'm dying* and then *Goodbye.*[36]

Mary got up and took a sleeping pill. Two hours later she woke up with tears running down her face. She asked herself:

Who decides which . . . deaths are tragic and which are not? Who decides what is big and what is little? Is it a matter of numbers or physical mass or intelligence? If you are a little creature or a little person dying alone and in pain, you may not remember or know that you are little. If you are in enough pain, you may not remember who or what you are; you may know only your suffering, which is immense . . . What decides – common sense? Can common sense dictate such things?[37]

A year after Gattino died, Mary was still looking for him. In the course of her search her feelings about human beings changed. Driving to a shelter to check whether he had been brought there, she heard a radio story about how American mercenaries shot dead a medical student in Iraq when he was getting out of his car, and then killed his mother when she jumped from the car to hold him. Previously she had heard such stories without feeling anything. Now they tore her heart open. 'It was the loss of the cat that had made this happen; his very smallness and lack of objective consequence had made the tearing open possible.'[38]

Her mind was also torn. She visited another psychic, who told her Gattino had died, probably of kidney failure from eating something toxic, and called another, who said he had died without suffering, curling up as if he were going to sleep. She put up another round of posters, and almost immediately got calls from people saying they had seen a small, one-eyed cat. Another security guard, 'a reticent older man', told her he had seen Gattino three months ago, but not since then. 'I haven't seen many cats lately,' he added. 'I'll tell you what I have seen, though. There's a huge bobcat, all over campus late at night. That and a lot of coyotes.' It was clear what he meant. Mary thought that at least it was a death an animal would understand.[39] Yet she felt he was still there. She dreamed of him for months. In the dream she would call him in the yard, and he would come to her as he had done in reality, 'running with his tail up, leaping slightly in his eagerness, leaping finally into my lap'.[40]

When her father was dying, Mary asked him a question. '"Daddy," I said, "tell me what you suffered. Tell me what it was like for you."' She didn't think he heard her, but she felt she heard part of an answer when she was out looking for Gattino late at night, when no one was around. 'It occurred to me then that the loss of my cat was in fact a merciful way for me to have my question answered.'[41] She was aware the feeling could be a result of magical thinking, but she was not convinced of this. What is real and what is imagined in human life is hard to decide:

> If someone had told me to smear shit on myself and roll in the yard, if that person was a cat expert and made a convincing case that, yes, doing so *could* result in the return of my cat, I probably would've done it. I did not consider this pathetic susceptibility 'magical thinking'. I didn't consider it very different from any other kind of thinking. It was more that the known, visible order of things had become unacceptable to me – senseless, actually – because it was too violently at odds with the needs of my disordered mind. Other kinds of order began to become visible to me, to bleed through

and knit together the broken order of what had previously been known. I still don't know if this cobbled reality was completely illusory, an act of desperate will, or if it was an inept and partial interpretation of something real, something bigger than what I could readily see.[42]

When Mary was giving up hope of finding Gattino, she went to Montana to do a reading at a university. Her hotel room overlooked a river, and one day she watched a dog being let off its chain and leap into the water, 'his legs splayed ecstatically wide'. She smiled and thought, 'Gattino'. Even if he was dead, he was there in that splayed, ecstatic leap. 'This idea was no doubt an illusion, a self-deception. But that dog was not. That dog was real. And so was Gattino.'[43]

Whether Gattino was still present in the world does not matter much. It is the fact that Gattino existed and what he did that are important. Mary's attachment to the cat was unlike any she had with human beings. The intertwined emotions of vanity and cruelty, remorse and regret that are at work in love between humans were absent. Her memories of Gattino changed her feelings towards her father, the children she hosted and the Iraqi medical student who had been shot. A love from beyond the human world untangled the love she had known with humans.

Among human beings love and hate are often mixed. We may love others deeply, and at the same time resent them. The love we feel for other human beings may become hateful to us, and be felt as a burden, a fetter on our freedom, while the love they feel for us can seem false and untrustworthy. If, despite these suspicions, we go on loving them, we may come to hate ourselves. The love animals may feel for us and we for them is not warped in these ways.

Losing Gattino was almost too hard for Mary to bear.

Yet his life was not sad in the way human lives can be sad. Gaitskill has written:

> To be human is finally to be a loser, for we are all fated to lose our carefully constructed sense of self, our physical strength, our health, our precious dignity and finally our lives.[44]

Gattino lived and died by chance, but he was not a loser. In his short, fearless, un-tragic life, he gave her something no human had been able to give her. For a while she was no longer ruled by laws of pleasure and pain. She no longer hated the people she loved – or herself for loving them. A tiny, one-eyed, seemingly inconsequential creature broke and remade her world. Perhaps Gattino was, after all, a magical animal.

5

Time, Death and the Feline Soul

Near the end of his autobiography, the Russian religious philosopher Nicolas Berdyaev writes of one of the most profound experiences in an eventful life:

> At the very time of the liberation of Paris we lost our beloved Muri, who died after a painful illness. His sufferings before death were to me the sufferings and travails of the whole creation; through him I was united to the whole creation and awaited its redemption. It was extremely moving to watch Muri, on the eve of his death, make his way with difficulty to Lydia's room (she was herself already seriously ill) and jump on her bed; he had come to say good-bye. I very rarely weep, but – this may sound strange or comic or trivial – when Muri died I wept bitterly. People speculate about 'the immortality of the soul', but there I was demanding immortal, eternal life for Muri. I would not have less than eternal life with him. A few months later I was to lose Lydia . . . I cannot be reconciled to death and the tragic finality of human existence . . . There can be no life unless it restores all those we love to itself.[1]

It is only if you have read the previous three hundred pages that you realize Muri was Berdyaev's cat. That the philosopher should have felt so much grief at Muri's passing may seem strange. But Berdyaev was no ordinary philosopher. Unlike most, then and now, he had seen a seemingly permanent human world pass away and vanish.

Born in Kiev in 1874, when Ukraine was part of the Russian empire, Berdyaev grew up as a solitary child in an aristocratic family. A freethinker, his father was sceptical of religion. Russian Orthodox by birth, his mother was critical of the established Church and leaned towards Catholicism. Throughout his life Berdyaev resisted any attempt to limit his freedom of thought. Following a family tradition he attended a military school, but soon left to study philosophy at the University of Kiev. Like many at the time he became a Marxist, and in 1898 was arrested at a demonstration and expelled from the university. Going on to work for an illegal underground press, he was arrested again and sentenced to three years of exile in Vologda. The conditions there were mild compared to those suffered by rebels against Tsarism, and incomparably less severe than in the camps that would be established by Lenin and Stalin.

When he returned to Kiev, Berdyaev met and married the poet Lydia Trusheff, with whom he shared the rest of his life, and the two moved to St Petersburg. No longer attracted by Marxism, but still a dissenter, he plunged into the intellectual life of the city during the years leading up to the First World War and the Russian Revolution. By now an overtly religious thinker, he published an article attacking the Holy Synod of the Orthodox Church for disciplining monks who deviated from official doctrine. Arrested on charges of blasphemy, he was sentenced to lifetime exile in Siberia, but the Bolshevik regime came to power and the punishment was never enforced.

Berdyaev soon clashed with the new regime. He was

allowed to lecture and write, and in 1920 was appointed professor of philosophy at the University of Moscow, but not long after he was arrested on charges of conspiracy and imprisoned. The much-feared head of Lenin's secret police Felix Dzerzhinsky visited him in his cell for an interrogation, which turned into a fiery exchange on Bolshevism. In September 1922 Berdyaev was expelled from the Soviet Union.

Along with other prominent members of the Russian intelligentsia – artists, scholars, scientists and writers – he left on what became known as 'the philosopher's steamship' – actually two vessels hired by the Bolshevik government to transport potentially troublesome intellectuals and their families to Germany. Others were sent by train to Riga in Latvia or by ship from Odessa to Istanbul. The plan of deporting the intelligentsia seems to have originated with Lenin himself.[2]

After arriving in Germany, Berdyaev and his wife moved first to Berlin, then to Paris, where they spent the rest of their lives. He was a prolific writer and participated in many dialogues with other Russian émigrés and the French intellectual community. He continued writing during the Nazi occupation, publishing the books after the war. He died at his desk in his home at Clamart, not far from Paris, in 1948.

The central questions that engaged Berdyaev had to do with time, death and eternity. He wrote:

> It has always surprised me how people can rely on the gradualness of human development, on the stability of human nature, on rational appeals to truth, on the objective standards of good, and all the other ambrosial illusions, in view of the unrelieved corruptibility and transitoriness of human life and the mortal wounds inflicted on man by every death, every parting, every betrayal, every passion.[3]

Berdyaev believed that, if death was the end, life had no sense. Life was a struggle for a meaning beyond life, which could redeem it from emptiness. What is unusual in Berdyaev is that he included his beloved cat in this struggle.

Whether Muri saw himself as joining in the search is doubtful. Lacking the human fear that death is the end of life's story, cats do not need another life in which the story continues. Yet Berdyaev's intuition that Muri sensed he was leaving the human beings with whom he had lived may have been well founded. Cats know when their life is coming to an end. As Doris Lessing found, they may also welcome its ending.

Describing how her black cat reacted when taken seriously ill, Lessing wrote:

> Her jaws and mouth were covered with white froth, a sticky foam which would not easily wipe off. I washed it off. She went back to the corner, crouching, looking in front of her. The way she sat was ominous: immobile, patient, and she was not asleep. She was waiting . . . cats decide to die. They creep into a cool place somewhere, because of the heat of their blood, crouch down, and wait to die.
>
> When I took black cat home [after a night at the cat hospital], she stalked gauntly into the garden. It was early autumn and cold. She crouched against the chill of the garden wall, cold earth under her, in the patient waiting position of the night before.
>
> I carried her in, and put her on a blanket, not too close to a radiator. She went back to the garden: same position, same deadly, patient position.
>
> I took her back and shut her in. She crept to the door, and settled down there, nose towards it, waiting to die.[4]

Lessing kept the cat indoors, and looked after her daily and hourly in the weeks that followed. The cat recovered, and some months later was the cat she had been, 'glossy, sleek, clean and purring'. She had forgotten she had been ill, but somewhere there remained in her mind a memory of the diagnostic room at the clinic, and she trembled, then froze, for hours when she had to be taken there again to treat an ear infection.

Lessing seems to have felt some guilt at what she had done to the cat by bullying her 'back into life against her will'. The cat, Lessing concludes, was 'a normal cat, with normal instincts'.[5]

CIVILIZATION AS DEATH-DENIAL

An idea of an afterlife emerged along with human beings. Around 115,000 years ago, graves were being fashioned containing animal bones, flowers, medicinal herbs and valuables such as ibex horns. By 35,000–40,000 years ago, complete survival kits – food, clothing and tools – were being placed in graves throughout the world.[6] Humankind is the death-defined animal.

As humans became more self-aware, the denial of death became more insistent. For the American cultural anthropologist and psychoanalytical theorist Ernest Becker (1924–74), the human flight from death has been the driving force of civilization. Fear of death is also the source of the ego, which humans build in order to shield themselves from helpless awareness of their passage through time to extinction.

More than most, Becker's life was formed by encounters with death. At the age of eighteen he joined the army and served in an infantry battalion that liberated a Nazi extermination camp. When he was dying of cancer in hospital in

December 1973, he told a visitor, the philosopher Sam Keen: 'You are catching me *in extremis*. This is a test of everything I've written about death. And I've got a chance to show how one dies.'[7] Becker's theories were set out in *The Denial of Death* (1973), for which he received a posthumous Pulitzer Prize in 1974, and developed further in *Escape from Evil*, which appeared two years after his death.

Of all the aspects of their condition from which humans strive to be diverted, death is the most threatening. Most cannot bear the thought of their own non-existence, and the more they try to forget it, the more it obsesses them. Rituals may enable them to leave this pain behind because they are practices engaging the entire organism and not only the mind. The way to escape anxiety is through what Becker calls 'the myth-ritual complex'. He writes:

> The myth-ritual complex is a social form for the channelling of obsessions . . . It automatically engineers safety and banishes despair by keeping people focussed on the noses in front of their faces. The defeat of despair is not mainly an intellectual problem for an active organism, but a problem of self-stimulation via movement. Beyond a given point man is not helped by more 'knowing', but only by living and doing in a partly self-forgetful way . . . Neurosis is the contriving of private obsessional ritual to replace the socially-agreed one now lost by the demise of traditional society. The customs and myths of the traditional society provided a whole interpretation of the meaning of life, ready-made for the individual; all he had to do was to accept living it as true. The modern neurotic must do just this if he is to be 'cured': he must welcome a living illusion.[8]

Here Becker distinguishes between traditional societies, in

which collective rituals relieve human beings from thoughts of death, and modern societies, where individuals are expected to deal with their anxieties by themselves. The reality of modern society is mass neurosis, repeatedly descending into outright insanity. Neuroses are not so much symptoms of illness as attempts as self-cure. The totalitarian movements of modern times are attempts of this kind. But human beings cannot cure themselves of being the kind of human being they have become – solitaries who remain alone even as they flee to the shelter of mass belonging.

Reasoning makes the modern neurosis worse:

> the characteristics the modern mind prides itself on are pre-cisely those of madness. There is no one more logical than the lunatic, more concerned with the minutiae of cause and effect. Madmen are the greatest reasoners we know, and that trait is one of the accompaniments of their undoing. All their vital processes are shrunken into the mind. What is the one thing they lack that sane men possess? The ability to be careless, to disregard appearances, to relax and laugh at the world. They can't unbend, can't gamble their whole existence, as did Pascal, on a fanciful wager. They can't do what religion has always asked: to believe in a justification of their lives that seems absurd.[9]

Human beings chase power in order to give themselves a sense of escaping death, and according to Becker human evil comes from the same impulse. The practice of cruelty serves to keep any thought of dying at bay:

> Sadism naturally absorbs the fear of death . . . because by actively manipulating and hating we keep our organism absorbed in the outside world; this keeps self-reflection and the fear of death in a state of low tension. We feel we are

masters over life and death when we hold the fate of others in our hands. As long as we can continue shooting, we think more of killing than of being killed. Or, as a wise gangster once put it in a movie, 'When killers stop killing they get killed.'[10]

As Becker points out, many modern ideologies have been immortality cults. Russian Bolshevism contained a powerful strand in which the conquest of mortality was the supreme goal of the revolution, and when Lenin was embalmed the goal of some of those involved was to resurrect him when scientific advance had developed the means to do so.[11] The project of defeating death through science has been revived in the West, with the Director of Engineering at Google, Ray Kurzweil, emerging as a prominent proponent of technological immortality.[12]

Becker's analysis is compelling. But human attitudes to death are contradictory, and not all religions and philosophies have been ways of denying mortality. In Greek polytheism, the gods are shown wearying of their freedom from death and envying humans their brief lives. When they intervene in the human world, they do so from boredom and to punish humans for their good fortune in being mortal. The oblivion that comes with death is one of the privileges of being human.

Other religions are ambiguous in their view of how to deal with being mortal. From one angle, Buddhism is an attempt to escape death. If you step off the wheel of rebirth, you will not have to die again. From another point of view, Buddhism is a quest for mortality.[13] Salvation means being free from the suffering of life. Once you are no longer reborn, you no longer have to suffer. But what if there is no transmigration of souls? After all, the Buddha taught that the soul is an illusion. If you are not reborn, you will be saved from suffering whatever you do. Your one and only death is total and final.

In this regard Epicurus has the advantage over Buddhism. If the aim is to end suffering, salvation is assured for every living thing since all will die. Yet Epicurus too was inconsistent. If humans want deliverance from suffering, they can end their lives as soon as they have the opportunity. Strangely, the ancient sage did not draw this conclusion and endorsed suicide only in extreme situations.

Human beings may struggle to be self-determining beings, as Spinoza suggested in his theory of *conatus*. But they may tire of the effort, and it is then that they may wish to end their lives. Short of self-destruction, many people have been drawn to philosophies that hold out the prospect of their disappearance as distinct individuals. They may involve merging with some metaphysical entity – the Platonic form of the good, or some kind of world-soul. Or they may be philosophies like Schopenhauer's, which promise to dissolve the self into nothingness.

Much of humankind finds being an individual a burden. Philosophies of history have been invented in order to lighten the load. Berdyaev knew that part of the appeal of communism was that it offered release from solitude, and today liberalism serves a similar need. If you are a separate soul, distinct and different from all others, your history and fate are your own. If, on the other hand, you are moving towards some kind of universal human oneness, you are no longer alone. Your life belongs in a larger story, a fable of collective human self-realization. Even if as an individual you die for ever, the meaning of your life is not lost.

But not all human beings fear dying, and some may want to die. A few wish they had not been born. Thwarted by the world, their *conatus* wants to cancel itself. They would be glad if their lives were erased completely.

Thomas Hardy pictured someone of this kind in his poem 'Tess's Lament'. The poem can be read as a commentary on Hardy's novel *Tess of the d'Urbervilles* (1891), the story of a country girl who struggles to assert herself against her circumstances and ends by being hanged for murdering her lover. Looking at her life, Tess would prefer it to be obliterated:

> It wears me out to think of it,
>   To think of it;
> I cannot bear my fate as writ,
>   I'd have my life unbe;
> Would turn my memory to a blot,
> Make every relic of me rot,
> My doings be as they were not,
>   And leave no trace of me![14]

Tess wants not to die but to vanish from the world as if she had never existed at all.

If cats could look back on their lives, might they wish that they had never lived? It is hard to think so. Not making stories of their lives, they cannot think of them as tragic or wish they had never been born. They accept life as a gift.

Humans are different. Unlike any other animal, they are ready to die for their beliefs. Monotheists and rationalists regard this as a mark of our superiority. It shows we live for the sake of ideas, not just instinctual satisfaction. But if humans are unique in dying for ideas, they are also alone in killing for them. Killing and dying for nonsensical ideas is how many human beings have made sense of their lives.

To identify yourself with an idea is to feel protected against death. Like the human beings who are possessed by them, ideas are born and die. While they may survive for generations, they still grow old and pass away. Yet, so long as they are in the grip of an idea, human beings are what Becker called 'living illusions'. By identifying themselves with an ephemeral fancy they can imagine they are out of time. By killing those who do not share their ideas, they can believe they have conquered death.

As predators, cats kill to live. Females are ready to die for their kittens, and cats regularly risk their lives to escape confinement. They differ from humans in that they do not kill and die in order to achieve any kind of immortality. There are no feline suicide warriors. When cats want to die it is because they no longer want to live.

Wittgenstein wrote:

> If by eternity is understood not endless temporal duration but timelessness, then he lives eternally who lives in the present.[15]

Because they think they can conceive the end of their lives, humans believe they know more of death than do other animals. But what human beings know as their oncoming death is an image generated in their minds by their awareness of passing time. Knowing only their lives as they live them, cats are mortal immortals that think of death only when it is nearly upon them. It is not hard to see how they came to be worshipped.

CATS AS GODS

Embodying a freedom and happiness that humans have never known, cats are strangers in the human world. If they have been seen as 'unnatural' creatures it is because they live according to their nature. Since no such life can be found among humans, cats came to be seen as demons or as gods.

To understand the worship of cats in ancient Egypt, you must set aside concepts that seem natural to us today. As Jaromir Malek has written:

> The division which we instinctively make between people and animals was not so strongly felt and the category 'animals' did not, in fact, exist. To put it differently, 'living beings' included gods, people, and animals. A theological treatise recorded under Shabako (716–702 BC) but perhaps composed as early as the third millennium BC, describes the heart and tongue of the creator-god Ptah being present in 'all gods, all people, all cattle, all worms, all that lives'. Just like people, animals were made by the creator god, worshipped him (in their own way) and were looked after by him. In certain exceptional cases, their link with the god may have even been more immediate than that of humans.[16]

Our way of thinking about archaic peoples is soaked in nineteenth-century myths of progress. In a pioneering history of ancient Egypt, John Romer has captured this mythology succinctly:

> The longer temporal narrative of the archaeological historians [of Egypt] . . . was a universal pseudo-evolutionary progress that ran straight from savagery through barbarism to the Ritz Hotel.[17]

In this rationalist myth ancient Egypt was a society given over to magical thinking. Unable to tell the difference between their own thoughts and the natural world, the people of this far-off time blurred the distinctions between life and death, gods and government. But this is to project back on to these ancient humans our own ideas and beliefs.

The ancient Egyptians had nothing like our modern conception of what it means to be human. Human beings were not unique in having a status in the world that other animals lacked. Later Greek and Roman ideas, in which the human mind came closest to a divine mind, were also absent, along with any idea of 'religion'. The modern separation of a sacred realm of worship from the 'secular' zone of everyday life did not exist. If you asked an ancient Egyptian what religion they followed, they would not understand you.

The idea of a supernatural realm, which is acquired from monotheism, did not exist either. The Egyptians inherited animist traditions in which the world was full of spirits. In these traditions humans were not superior to other animals. There were not two wholly distinct orders of things – one of insentient matter and another of immaterial souls – but one that animal and human souls shared in common. Many of our most fundamental and seemingly self-evident categories of thought were absent.

In the philosophies of the past few centuries, human civilization advances in a majestic march that leads triumphantly to ourselves. Archaic minds are supplanted by modern ones. Myths and rituals yield to scientific explanation and pragmatic reasoning. Any idea that cats are magical animals must be part of a primitive past.

And yet the human mind has not changed much since archaic times, and the idea that we are quite different from the ancient Egyptians is itself rather primitive. We know far more than they did, and exercise far greater power over aspects of the material world, but that does not make us any less given to myth-making.

Once a dated mythology of progress has been left behind, a different view of the worship of cats emerges. Cats became gods in ancient Egypt by a natural process. They began interacting and then living with humans much as they did in the Near East.

Around 4000 BC wild cats strayed into Egyptian settlements and found granaries containing rodents and snakes, which they killed and ate. Over the next 2,000 years a symbiosis developed, with cats benefiting from dependable food supplies and humans from a reduction in vermin. From 2000 BC onwards cats introduced themselves into households and were accepted as companions. 'In this way,' Malek writes, 'the cat eventually became a domesticated animal, or to put it more precisely, domesticated itself.'[18]

In a small tomb at Abydos, a Middle Kingdom cemetery in Upper Egypt dated somewhere around 1980–1801 BC, seventeen skeletons of cats were found near a row of small pots which may have originally contained milk. If so, this would be the earliest recorded example of adult cats being fed in this way.[19] Between 1000 BC and AD 350, cats came to be regarded as manifestations of deities, in particular the goddess Bastet, and were being bred in temple catteries. In a 1250 BC a stela (a small round-topped slab of stone often placed at the back of large temples) of two cats is shown representing Pre (Ra, the sun-god). The stela contains a poem that seems addressed at once to the 'great cat' and to the sun-god:

> Giving praise to the great cat,
> kissing the earth before Pre, the great god:
> O peaceful one, who returns to peace,
> you cause me to see the darkness of your making.
> Lighten me that I can perceive your beauty,
> turn towards me,
> O beautiful one when at peace,
> the peaceful one who knows a return to peace.[20]

From being household helpers and companions, cats became omens of good fortune and sacred animals. Amulets showing cats were worn on the body or on clothing. By the time of the New Kingdom (after 1540 BC) cats were shown in royal tombs guarding the sun-god in his nightly passage through the underworld. In 'books of the afterlife' from this period, cats are portrayed watching over enemies of the god and standing as sentries at the last gate through which he must pass in his journey back to life and light. Statuettes showed cats in the company of gods, supporting or guarding them. At times human beings appear, kneeling in worship before cats themselves.

By the fourth century BC, a 'temple of the living cat' existed in the necropolis of Hermopolis, with a large cemetery of mummified cats nearby. Cats were not alone in being mummified; so, too, were mongooses, ibises, vultures, hawks and crocodiles, for example, and of course humans. But cats were mummified in vast numbers, and towards the end of the nineteenth century boatloads of these mummies were shipped to Europe. As there was a glut in the market, cat mummies were often used as fertilizer or even as ballast for ships, with many being destroyed or lost.

Herodotus writes that when an Egyptian house was on fire, the inhabitants were more concerned about their cats than their property. When a member of a visiting Roman delegation killed a cat accidentally in 59 BC, the man was lynched despite intervention from the king. And the Egyptian sage Ankhsheshonq warned, 'Do not laugh at a cat.'[21]

Cats have had a bad reputation among monotheists: the second-century Christian theologian Clement of Alexandria was already attacking the Egyptians for having cats in their temples. But some theistic traditions have been more respectful: the Italian Catholic friar St Francis of Assisi (1182–1226) believed that love of God's creation included love for all of God's creatures; Jewish law contains commandments requiring that animals be treated with compassion, including a 3,000-year-old injunction that farm animals be given a day of rest. The Prophet Muhammad is reputed to have cut off a sleeve so that he would not disturb a cat sleeping on it, while the medieval sultan Baibars (c.AD 1223–77) bequeathed a garden as a retreat for homeless cats in Cairo.

Cats were many things in ancient Egypt: sometimes companions of human beings as they passed on to another life, at other times manifestations of gods, at still others protectors of the gods. That they could be all of these at once testifies to the subtlety of the archaic Egyptian mind. But it also speaks to the presence of cats themselves. Cats symbolized an affirmation of life in a world preoccupied with the dead. Egyptian religion responded to the prospect of death by preparing for life in another world, but it needed cats to preserve a sensation of being alive in the realm beyond the grave. Knowing only life until they are on the brink of dying, cats are not ruled by death. The Egyptians had good reason for wanting cats to join them in the journey through the underworld.

When it came to death, humans and cats were in the same boat. No one in ancient Egypt believed that humans have souls while cats do not. But if the soul is untouched by death, the feline soul is closer to immortality than the human soul can ever be.

6

Cats and the Meaning of Life

If cats could understand the human search for meaning they would purr with delight at its absurdity. Life as the cat they happen to be is meaning enough for them. Humans, on the other hand, cannot help looking for meaning beyond their lives.

The search for meaning comes with awareness of death, which is a product of human self-consciousness. Fearing their lives ending, human beings invented religions and philosophies in which the meaning of their lives carried on after them. But the meaning humans make is easily broken, so they live in greater fear than before. The stories they have fashioned for themselves take over, and they spend their days trying to be the character they have invented. Their lives belong not to them but to a figure conjured up in their imagination.

One consequence of this way of living is that human beings may become fixated on occasions when their stories are disrupted. They may lose loved ones, find their own lives in danger or be forced to leave their homes. Those who turn their life into a tragic story are coping with experiences of irremediable loss. But it is a way of coping that comes with a cost. While thinking of your life as a tragedy may give it meaning, it binds you to your sorrows.

Cats may endure terrible suffering, and their lives may be brutally cut short. Mèo's life contained many horrors and when traumatic memories were triggered, they would return to him. Gattino suffered at the start and very likely the end of his life. Both cats knew much pain, neither of them anything of tragedy. Despite suffering, they lived with fearless joy. Can humans live like this? Or is humankind too frail for such a life?

CAT NATURE, HUMAN NATURE

There are many who would like to delete the idea of human nature from the lexicon. Human beings create themselves, they say. Unlike other animals, they can choose to be whatever they want to be. Talk of human nature is a way of curbing this freedom and leaving human beings to be ruled by the power of arbitrary norms.

This is called post-modernism and is promoted by thinkers such as Jean Baudrillard and Richard Rorty – it has had many incarnations. As preached by the early Jean-Paul Sartre, existentialism was the idea that humans have no nature, only histories they have fashioned for themselves. The Romantics wanted each human life to be a work of art, created – like the best artworks, they believed – from nothing. But if humans are like other living things in being the random spawn of evolution, how could they create themselves? It is true that the human animal fashions an artificial nature for itself. This is part of what Pascal means when he writes: 'Habit is a second nature that destroys the first. But what is nature? Why is habit not natural? I am very much afraid that nature itself is only a first habit, just as habit is a second nature.'[1]

But this second nature may be more superficial than Pascal believed.

The Russian writer Varlam Shalamov, who survived for fifteen years in Arctic Gulag camps where winter temperatures routinely fell below 50 degrees Centigrade and the average lifespan was around three years, observed that a few weeks of extreme cold, hunger, overwork and beatings were enough to destroy the humanity in any human being. Aside from isolated examples of kindness, there is nothing in Shalamov's account of the resilience of 'the human spirit'. Only non-human creatures show goodwill: the bears and the bullfinch that drew the hunter's fire so their mates could escape, the husky that protected prisoners and growled at the guards and the cat that helped inmates catch fish.

Human beings quickly lose their humanity, whereas cats never stop being cats. But if the nature human beings believe they have is composed of habits that can crumble away in weeks, what is there in human beings that is truly their own?

Contrary to the post-modernists, there is such a thing as human nature. It is expressed in the universal demand for meaning, for one thing. But human nature has produced many divergent and at times antagonistic forms of life. How can anyone know their own nature, when human nature is so contradictory? Might the idea that each of us has a nature of our own be just another metaphysical fiction?

The truth in the fiction of an individual nature is that the good life for each of us is not chosen but found. Even when they come from decisions we believe we have made, our experiences are not determined by us. The good life is not the life you want but one in which you are fulfilled. Stripped of metaphysics, this is Spinoza's idea of *conatus* and the Taoist belief that we must follow the way within us.

In this we are at one with all other creatures. Humans do not rank above other animals, or below them. There is no cosmic scale of value, no great chain of being; no external standard by which the worth of a life can be judged. Humans are humans, cats are cats. The difference is that, while cats have nothing to learn from us, we can learn from them how to lighten the load that comes with being human.

One burden we can give up is the idea that there could be a perfect life. It is not that our lives are inevitably imperfect. They are richer than any idea of perfection. The good life is not a life you might have led or may yet lead, but the life you already have. Here, cats can be our teachers, for they do not miss the lives they have not lived.

TEN FELINE HINTS ON HOW TO LIVE WELL

Cats have no interest in teaching humans how to live, and if they did they would not do so by issuing commandments. Yet one can imagine cats could give us hints as to how to live less awkwardly. Obviously, they would not expect us to apply their advice. They would offer their suggestions playfully, as a form of entertainment for themselves and the human beings who received them.

1 Never try to persuade human beings to be reasonable

Trying to persuade human beings to be rational is like trying to teach cats to be vegans. Human beings use reason to bolster whatever they want to believe, seldom to find out if what they believe is true. This may be unfortunate, but there is nothing you or anyone else can do about it. If human unreason frustrates or endangers you, walk away.

2 It is foolish to complain that you do not have enough time

If you think you do not have enough time, you do not know how to pass your time. Do what serves a purpose of yours and what you enjoy doing for its own sake. Live like this, and you will have plenty of time.

3 Do not look for meaning in your suffering

If you are unhappy, you may seek comfort in your misery, but you risk making it the meaning of your life. Do not become attached to your suffering, and avoid those who do.

4 It is better to be indifferent to others than to feel you have to love them

Few ideals have been more harmful than that of universal love. Better cultivate indifference, which may turn into kindness.

5 Forget about pursuing happiness, and you may find it

You will not find happiness by chasing after it, since you do not know what will make you happy. Instead, do what you find most interesting and you will be happy knowing nothing of happiness.

6 Life is not a story

If you think of your life as a story, you will be tempted to write it to the end. But you do not know how your life will end, or what will happen before it does. It would be better to throw the script away. The unwritten life is more worth living than any story you can invent.

7 Do not fear the dark, for much that is precious is found in the night

You have been taught to think before you act, and often that may be good advice. Acting on how you feel at the moment may be no more than obeying worn-out philosophies you have accepted without thinking. But sometimes it is better to follow an inkling that glimmers in the shadows. You never know where it may lead you.

8 Sleep for the joy of sleeping

Sleeping so that you can work harder when you wake up is a miserable way to live. Sleep for pleasure, not profit.

9 Beware anyone who offers to make you happy

Those who offer to make you happy do so in order that they themselves may be less unhappy. Your suffering is necessary to them, since without it they would have less reason for living. Mistrust people who say they live for others.

10 If you cannot learn to live a little more like a cat, return without regret to the human world of diversion

Living like a cat means wanting nothing beyond the life you lead. This means living without consolations, and that might be too much for you to bear. If so, take up an old-fashioned religion, preferably one that abounds in rituals. If you cannot find a faith that suits you, lose yourself in common life. The excitement and disappointments of romantic love, the pursuit of money and ambition, the charades of politics and the clamour of the news will soon banish any sense of emptiness.

MÈO ON THE WINDOW LEDGE

A feline philosopher would not encourage humans to seek wisdom. If you do not take pleasure in life itself, find fulfilment in inconstancy and illusion. Do not struggle against fears of death. Let them die away. If you crave tranquillity you will be forever in turmoil. Instead of turning away from the world, turn back to it and embrace its folly.

At times you may want to return to yourself. Looking at the world without struggling to fit it into our stories is what many traditions call contemplation. When you see things without wanting to change them, they can give you a glimpse of eternity. Each moment is complete, and the shifting scene reveals itself to you as if it were out of time. Eternity is not another order of things but the world seen without anxiety.

For humans, contemplation is a break from living; for cats it is the sensation of life itself. Mèo lived always in danger, and spent many hours perched precariously on the window ledge. He did not search for meaning in the world he watched below. Cats show us that seeking after meaning is like the quest for happiness, a distraction. The meaning of life is a touch, a scent, which comes by chance and is gone before you know it.

Acknowledgements

My editor at Penguin, Simon Winder, has given me unfailing encouragement. His and his colleague Eva Hodgkin's comments have improved the text immeasurably. My agent at the Wylie Agency, Tracy Bohan, and her colleague Jennifer Bernstein have been wonderfully supportive and helpful from the book's conception. Adam Phillips has stirred my thoughts over many years on the themes the book pursues and his comments have been invaluable. Conversations with Bryan Appleyard, Robert Colls, Michael Lind, Paul Schutze, Will Self, Geoffrey Smith, Sheila Stevens and Marina Vaizey helped me write it.

Four cats made their own indispensable contribution. Two Burmese sisters, Sophie and Sarah, and two Birman brothers, Jamie and Julian, were cherished companions over a period of nearly thirty years. Julian was in his twenty-third year, still enjoying life, while I was writing the book.

As ever my deepest gratitude goes to my wife, Mieko, without whom none of this would have happened.

John Gray

Notes

I CATS AND PHILOSOPHY

1 I have discussed this rationalist view of religion in *Seven Types of Atheism* (London: Penguin Books, 2019), pp. 9–14.

2 Arthur Schopenhauer, *The World as Will and Representation*, vol. 2, translated by E. F. J. Payne (New York: Dover Publications, 1966), pp. 482–3.

3 See Peter Godfrey-Smith, *Other Minds: The Octopus and the Evolution of Intelligent Life* (London: William Collins, 2017), Chapter 4, 'From White Noise to Consciousness', pp. 77–105.

4 I consider ideas of cosmic evolution in *The Immortalization Commission: The Strange Quest to Cheat Death* (London: Penguin Books, 2012), pp. 213–19.

5 For the view that humans may be the only conscious beings in the cosmos, see James Lovelock, *Novacene: The Coming Age of Hyperintelligence* (London: Allen Lane, 2019), pp. 3–5.

6 Michel de Montaigne, *An Apology for Raymond Sebond*, translated and edited by M. A. Screech (London: Penguin Books, 1993), p. 17.

7 Montaigne, *Apology for Raymond Sebond*, pp. 16, 17.

8 Sextus Empiricus, *Outlines of Scepticism*, edited by Julia Annas and Jonathan Barnes (Cambridge: Cambridge University Press, 2000), pp. 5–6.

9 Montaigne, *Apology for Raymond Sebond*, p. 53.

10 Montaigne, *Apology for Raymond Sebond*, p. 54.

11 For Wittgenstein's idea of a homoeopathic philosophy/anti-philosophy, see K. T. Fann, *Wittgenstein's Conception of Philosophy* (Singapore: Partridge Publishing, 2015). In an Appendix, Fann explores some affinities between Wittgenstein's later work and Taoism (see pp. 99–114). Montaigne's scepticism in regard to philosophy is explained in Hugo Friedrich, *Montaigne*, edited with an introduction by Philippe Desan, translated by Dawn Eng (Berkeley, CA: University of California Press, 1991), pp. 301–9.

12 John Laurence, *The Cat from Hué: A Vietnam War Story* (New York: PublicAffairs, 2002), p. 23.

13 Laurence, *The Cat from Hué*, p. 496.

14 Laurence, *The Cat from Hué*, p. 489.

15 Laurence, *The Cat from Hué*, p. 485.

16 Laurence, *The Cat from Hué*, pp. 491, 498-9.

17 Laurence, *The Cat from Hué*, p. 498.

18 Laurence, *The Cat from Hué*, p. 820.

19 Laurence, *The Cat from Hué*, p. 822.

20 Laurence, *The Cat from Hué*, p. 822.

21 For an authoritative account of the domestication of cats, see Abigail Tucker's *The Lion in the Living Room: How House Cats Tamed Us and Took over the World* (New York and London: Simon and Schuster, 2016), pp. 31-5.

22 Tucker, *The Lion in the Living Room*, p. 32.

23 Tucker, *The Lion in the Living Room*, p. 47.

24 Elizabeth Marshall Thomas, *The Tribe of Tiger: Cats and Their Culture*, illustrated by Jared Taylor Williams (London: Orion Books, 1995), p. 3.

25 See Peter P. Marra and Chris Santella, *Cat Wars: The Devastating Consequences of a Cuddly Killer* (Princeton, NJ: Princeton University Press, 2016), p. 19.

26 Carl Van Vechten, *The Tiger in the House* (New York: Dover Publications, 1996), p. 75.

27 Keith Thomas, *Man and the Natural World: Changing Attitudes in England 1500-1800* (London: Allen Lane, 1983), pp. 109-10.

28 Robert Darnton, *The Great Cat Massacre and Other Episodes in French Cultural History* (New York: Basic Books, 2009), p. 96.

29 Van Vechten, *The Tiger in the House*, pp. 74-5.

2 WHY CATS DO NOT STRUGGLE TO BE HAPPY

1 George Santayana, *Three Philosophical Poets: Lucretius, Dante, Goethe* (New York: Doubleday, Anchor Books, 1953), p. 183.

2 Marcus Aurelius, *Meditations*, translated by A. S. L. Farquharson (Oxford: Oxford University Press, 2008), p. 13.

3 Joseph Brodsky, 'Homage to Marcus Aurelius', in Joseph Brodsky, *On Grief and Reason: Essays* (London: Penguin Books, 2011), p. 245.

4 Seneca, *Epistles 66–92*, translated by Richard M. Gummere (Cambridge, MA, and London: Harvard University Press, 2006), pp. 177, 179, 181.

5 Blaise Pascal, *Pensées*, translated with an introduction by A. J. Krailsheimer (London: Penguin Books, 1966), p. 66.

6 Pascal, *Pensées*, pp. 67–8.

7 Pascal, *Pensées*, pp. 39, 41.

8 Michel de Montaigne, 'On diversion', in Michel de Montaigne, *The Complete Essays*, translated by M. A. Screech (London: Penguin Books, 2003), p. 941.

9 Montaigne, *Complete Essays*, 'On affectionate relationships', pp. 205–19.

10 Pascal, *Pensées*, p. 59.

11 Pascal, *Pensées*, 'The Memorial', pp. 309–10.

12 Pascal, *Pensées*, p. 60.

13 Pascal, *Pensées*, p. 44.

14 For Pascal's Wager, see *Pensées*, pp. 149–55.

15 Pascal, *Pensées*, p. 274.

16 Pascal, *Pensées*, p. 95.

17 James Boswell, *Life of Johnson*, edited by R. W. Chapman (Oxford: Oxford University Press, 1980), p. 368.

18 Samuel Johnson, *The History of Rasselas, Prince of Abissinia*, edited by Thomas Keymer (Oxford: Oxford University Press, 2009), p. 42.

19 Christopher Smart, 'For I will consider my Cat Jeoffry'. Often anthologized and available in *The Sophisticated Cat*, edited by Joyce Carol Oates and Daniel Halpern (London: Pan Books, 1994), pp. 61–4.

20 Johnson, *The History of Rasselas*, p. 93.

3 FELINE ETHICS

1 Pascal, *Pensées*, p. 47.

2 See Alasdair MacIntyre, *After Virtue: A Study in Moral Theory*, 3rd edn (London: Bloomsbury Academic, 2007), pp. 27–41.

3 See Bernard Williams, *Ethics and the Limits of Philosophy* (London: Routledge, 2011), Chapter 10, 'Morality, the Peculiar Institution', pp. 193–218.

4 Aristotle, *History of Animals*, translated by D'Arcy Wentworth Thompson (Whitefish, MT: Kessinger Publishers, 2004).

5 For the good life among dolphins, see Alasdair MacIntyre, *Dependent Rational Animals: Why Human Beings Need the Virtues* (London: Duckworth, 1999), pp. 23–6.

6 See A. C. Graham, *Disputers of the Tao: Philosophical Argument in Ancient China* (La Salle, IL: Open Court, 1989), pp. 13–14, 191–2.

7 Darwin's failure to stick consistently to his theory of natural selection as a purposeless process is discussed in my book *Seven Types of Atheism* (London: Penguin Books, 2019), pp. 54–5.

8 See my book *Straw Dogs: Thoughts on Humans and Other Animals* (London: Granta Books, 2002).

9 Antonio Damasio, *Looking for Spinoza* (London: Vintage Books, 2004), pp. 170–71. For an illuminating discussion of mind/body unity, see also Damasio's *Self Comes to Mind: Constructing the Conscious Brain* (New York, Pantheon Books, 2010).

10 Stuart Hampshire, 'Spinoza and the Idea of Freedom', in *Spinoza: A Collection of Critical Essays*, edited by Marjorie Grene (Garden City, NY: Anchor Press/Doubleday, 1973), pp. 303–4. Reprinted in Stuart Hampshire, *Spinoza and Spinozism* (Oxford: Clarendon Press, 2005), pp. 182–4.

11 Hampshire, 'Spinoza and the Idea of Freedom', p. 312.

12 See Daniel M. Wegner, *The Illusion of Conscious Will* (London: MIT Press, 2002).

13 Hampshire, *Spinoza and Spinozism*, p. 13.

14 Hampshire, *Spinoza and Spinozism*, p. 13.

15 Benedict Spinoza, *Ethics; and Treatise on the Correction of the Intellect*, translated by Andrew Boyle, revised by, and with an introduction and notes by, G. H. R. Parkinson (London: J. M. Dent, 1993), pp. 172–3.

16 Thomas Hobbes, *Leviathan*, edited with an introduction and notes by J. C. A. Gaskin (Oxford: Oxford University Press, 2008), p. 66.

17 Spinoza, *Ethics; and Treatise on the Correction of the Intellect*, p. 89.

18 Spinoza, *Ethics; and Treatise on the Correction of the Intellect*, p. 183.

19 See Stephen Lukashevich, *Konstantin Leontev (1831–1891): A Study in Russian 'Heroic Vitalism'* (New York: Pageant Press, 1967), Chapter V.
20 I have criticized theories of effective altruism in 'How & How Not to Be Good', *New York Review of Books*, 21 May 2015, reprinted as 'How Not to Be Good: Peter Singer on Altruism', in *Gray's Anatomy: Selected Writings*, new edition (London: Penguin Books, 2016), pp. 482–91.
21 Philip Kitcher, *The Ethical Project* (Cambridge, MA: Harvard University Press, 2011), p. 7.
22 Paul Wienpahl, *The Radical Spinoza* (New York: New York University Press, 1979), pp. 89–90.
23 Jon Wetlesen, *The Sage and the Way: Spinoza's Ethics of Freedom* (Assen: Van Gorcum, 1979), p. 317.
24 Atsuko Saito, Kazutaka Shinozuka, Yuki Ito and Toshikazu Hasegawa, 'Domestic cats (*Felis catus*) discriminate their names from other words', *Scientific Reports* 9 (5394), 4 April 2019.
25 For an illuminating discussion of attention and distraction, see Adam Phillips, *Attention Seeking* (London: Penguin Books, 2019).
26 See Eugen Herrigel, *Zen in the Art of Archery: Training the Mind and Body to Become One*, translated by R. F. C. Hull (London: Penguin Books, 2004).

4 HUMAN VS FELINE LOVE

1 Judith Thurman, *Secrets of the Flesh: A Life of Colette* (London: Bloomsbury, 1999), p. 397.
2 Colette, 'The Cat', in Colette, *Gigi and The Cat*, translated by Roger Senhouse (London: Vintage Books, 2001), p. 108.
3 Colette, 'The Cat', p. 155.
4 Colette, 'The Cat', p. 157.
5 J. R. Ackerley, *My Dog Tulip* (New York: New York Review of Books, 2011).
6 Andrew Wilson, *Beautiful Shadow: A Life of Patricia Highsmith* (London: Bloomsbury, 2003), p. 333.
7 For some profound observations on feline attachment, see Jeffrey Masson, *The Nine Emotional Lives of Cats: a Journey into the Feline Heart* (London: Vintage, 2003), 53–59.
8 Wilson, *Beautiful Shadow*, pp. 331, 332, 267.

9 Wilson, *Beautiful Shadow*, p. 331.
10 Wilson, *Beautiful Shadow*, p. 331.
11 Wilson, *Beautiful Shadow*, p. 331.
12 Patricia Highsmith, 'Ming's Biggest Prey', in her *The Animal-Lover's Book of Beastly Murder* (London: Penguin Books, 1979), pp. 57–8.
13 Highsmith, 'Ming's Biggest Prey', p. 67.
14 Highsmith, 'Ming's Biggest Prey', p. 68.
15 See Patricia Highsmith, *Zeichnungen* (Zurich: Diogenes, 1995).
16 Germaine Brée, *Marcel Proust and Deliverance from Time* (London: Chatto and Windus, 1956), pp. 99–100.
17 Junichirō Tanizaki, *In Praise of Shadows*, translated by Thomas J. Harper and Edward G. Seidensticker (London: Vintage Books, 2001), p. 46.
18 Tanizaki, *In Praise of Shadows*, p. 20.
19 Tanizaki, *In Praise of Shadows*, p. 20.
20 Junichirō Tanizaki, *A Cat, a Man, and Two Women*, translated by Paul McCarthy (London: Daunt Books, 2017), pp. 4–5.
21 Tanizaki, *A Cat, a Man, and Two Women*, pp. 103–4.
22 Tanizaki, *A Cat, a Man, and Two Women*, p. 120.
23 Mary Gaitskill's memoir appeared first in *Granta* magazine in 2009, issue 107, and was republished in her essay collection *Somebody with a Little Hammer* (New York: Vintage Books, 2018), pp. 131–79.
24 See Parul Sehgal, 'Mary Gaitskill and the Life Unseen', *The New York Times*, 2 November 2015.
25 Mary Gaitskill, *This is Pleasure* (London: Serpent's Tail, 2019).
26 Gaitskill, 'Lost Cat: A Memoir', in *Somebody with a Little Hammer*, p. 134.
27 Gaitskill, 'Lost Cat', p. 131.
28 Gaitskill, 'Lost Cat', p. 135.
29 Gaitskill, 'Lost Cat', pp. 135–6.
30 Gaitskill, 'Lost Cat', pp. 136–7.
31 Gaitskill, 'Lost Cat', p. 137.
32 Gaitskill, 'Lost Cat', p. 138.
33 Gaitskill, 'Lost Cat', p. 137.
34 Gaitskill, 'Lost Cat', p. 138.
35 Gaitskill, 'Lost Cat', p. 146.
36 Gaitskill, 'Lost Cat', pp. 149–51.
37 Gaitskill, 'Lost Cat', p. 151.
38 Gaitskill, 'Lost Cat', p. 154.
39 Giatskill, 'Lost Cat', p. 173.

40 Gaitskill, 'Lost Cat', p. 171.
41 Gaitskill, 'Lost Cat', p. 158.
42 Gaitskill, 'Lost Cat', pp. 162–3.
43 Gaitskill, 'Lost Cat', p. 179.
44 Mary Gaitskill, 'Victims and Losers: A Love Story', in *Somebody with a Little Hammer*, p. 82.

5 TIME, DEATH AND THE FELINE SOUL

1 Nicolas Berdyaev, *Self-Knowledge: An Essay in Autobiography*, translated by Katharine Lampert (San Rafael, CA: Semantron Press, 2009), pp. 319–20, 323.
2 For a vivid account of Lenin's deportation of the Russian intelligentsia, see Lesley Chamberlain, *The Philosophy Steamer: Lenin and the Exile of the Intelligentsia* (London, Atlantic Books, 2006).
3 Berdyaev, *Self-Knowledge*, pp. 291–2.
4 Doris Lessing, *On Cats* (London: HarperCollins, 2008), pp. 86–7.
5 Lessing, *On Cats*, pp. 97–8.
6 Felipe Fernández-Armesto, *Out of Our Minds: What We Think and How We Came to Think It* (London: Oneworld Publications, 2019), pp. 35–7.
7 Sam Keen's long conversation with Ernest Becker was published as 'The heroics of everyday life: a theorist of death confronts his own end', *Psychology Today*, April 1974.
8 Ernest Becker, *The Denial of Death* (London: Souvenir Press, 2011, reprinted 2018), p. 199.
9 Becker, *The Denial of Death*, p. 201.
10 Ernest Becker, *Escape from Evil* (New York: The Free Press, 1975), pp. 113–14.
11 For a detailed account of Bolshevism as an immortality ideology, see my book *The Immortalization Commission: The Strange Quest to Cheat Death* (London: Penguin Books, 2012).
12 See Gray, *The Immortalization Commission*, pp. 213–16.
13 I discuss the quest for mortality in Buddhism in *Straw Dogs: Thoughts on Humans and Other Animals* (London, Granta Books, 2002), pp. 129–30.
14 'Tess's Lament', in *Thomas Hardy: Selected Poetry*, edited with an introduction and notes by Samuel Hynes (Oxford: Oxford University Press, 1996), p. 40.

15 Ludwig Wittgenstein, *Tractatus Logico-Philosophicus*, translated by C. K. Ogden, with an introduction by Bertrand Russell (New York: Dover Publications, 1999), section 6.4311, p. 106.

16 Jaromir Malek, *The Cat in Ancient Egypt* (London: British Museum Press, 2017), pp. 75–6.

17 John Romer, *A History of Ancient Egypt from the First Farmers to the Great Pyramid* (London: Penguin Books, 2013), p. xix.

18 Malek, *The Cat in Ancient Egypt*, p. 55.

19 Malek, *The Cat in Ancient Egypt*, p. 51.

20 Malek, *The Cat in Ancient Egypt*, p. 89.

21 Malek, *The Cat in Ancient Egypt*, pp. 75, 100.

6 CATS AND THE MEANING OF LIFE

1 Blaise Pascal, *Pensées*, translated with an introduction by A. J. Krailsheimer (London, Penguin Books, 1966), p. 61.